AF749303

CONTRIBUTIONS
TO
ASIAN STUDIES

A semi-annual publication sponsored by the Canadian Association for South Asian Studies

Contributions to Asian Studies, a publication sponsored by the Canadian Association for South Asian Studies, is published twice a year. Articles in the *CAS* represent neither the views of the Association nor those of the editors.

Editorial: Manuscripts and communications to the editor should be addressed to: K Ishwaran, Department of Sociology and Anthropology, York University, 4700 Keele Street, Toronto, Ontario M3J 1P3.

Annual Membership Dues of the Association, beginning every calendar year, are $6.00. Institutional membership, including *Contributions to Asian Studies*, $ 50.00 annually. Application for membership and payment of dues should be made in the name of the Canadian Association for South Asian Studies and mailed to Dr. M. D. Chaudhry, Department of Political and Economic Science, Royal Military College, Kingston, Ontario K7L 2W3.

Newsletter: CASAS also publishes periodically newsletters which are now edited by Parvez A. Wakil, University of Saskatchewan, Saskatoon, and Dr. M. Fayyaz, Queen's University, Kingston, Ontario. All correspondence relating to the newsletter should be addressed to them.

Subscription and All Other Business Correspondence should be addressed to the publisher:
E. J. Brill, Leiden, The Netherlands

Since the size of each volume varies, the price of each volume is not determined in advance. Members of CASAS are allowed a 10% discount.

ISBN 90 04 04529 5

PRINTED IN THE NETHERLANDS

CONTRIBUTIONS
ASIAN STUDIES

General Editor: K. Ishwaran

VOLUME 9

POPULATION LAND AND STRUCTURAL CHANGE IN SRI LANKA AND THAILAND

EDITOR: JAMES BROW

LEIDEN
E. J. BRILL
1976

Contents

An Introduction to Land, Population and Structure: Three Guises of the Man-Land Ratio 1
L. M. Hanks

The Closing of the Frontier: Land Pressures and their Implications for Rural Social Organization in the Thai Central Plain 7
Steven Piker

Opening a Frontier: The Expansion of Rice Cultivation in Central Thailand in the 1890's 27
David B. Johnston

In Search of Land: Village Formation in the Central Chi River Valley, Northeastern Thailand 45
Charles F. Keyes

Land Use and Population Growth in Colonial Ceylon 64
Patrick Peebles

The Impact of Population Growth on the Agricultural Practices and Settlement Patterns of the Anuradhapura Veddahs 80
James Brow

Kinship and Land Rights in the Context of Demographic Change . . . 97
A. J. Selvadurai

Contributors 113

CONTRIBUTIONS TO ASIAN STUDIES SERIES

Despite socio-cultural, economic, political and geographical diversity, the countries of Asia display certain broad similarities. In general, all of them are products of ancient civilizations, faced with the complex forces of modernity, which they have responded to in quite complex ways. Asia's unity in diversity, the dynamics of its social and cultural patterns, past and present, have stimulated a large body of scholarly work. A new phase in Asian studies has developed through first rate contributions from indigenous scholars who often bring an outlook and understanding of their own. *Contributions to Asian Studies*, a semi-annual publication, is intended as a forum for scholarly analyses of Asian societies and cultures, past and contemporary, from the diverse standpoints of the international community of scholars in all the social sciences and humanities.

EDITORIAL NOTE

Earlier versions of four of the articles printed here, those of Steven Piker, Charles F. Keyes, A. J. Selvadurai and James Brow, were first presented at a symposium on "Population, Land and Structural Change in Sri Lanka and Thailand", held at the annual meeting of the American Anthropological Association in Mexico City in November 1974. Lucien Hanks, who contributes an introduction to the present collection, was the discussant at the Mexico City symposium. The symposium was organized by Steven Piker and James Brow.

An Introduction to Land, Population and Structure: Three Guises of the Man-Land Ratio

L. M. HANKS

North Bennington, Vermont

THE NIGHTMARE death of a social order has been played out so often by prophets, actors and poets as well as in history itself, that this spectre is easily aroused from cultural memory. Plague, fire, war, drought, flood and the angel horsemen of the Apocalypse do not exhaust the litany of disasters descending to us from the ancients. Nonetheless, early in the 19th century a new spectre appeared when Thomas Robert Malthus, ruminating on the perfectibility of human society, envisioned an uncontrollable increase of population. Since then the prospect of overpopulation, insufficient food, and a final day of violence has not ceased to provoke discussion. Is population perfectible by eugenic selection? Can its growth be limited by public campaign? Is tampering with life sinful? How can the world feed all its people?

A 19th century phrasing of the problem, having today a slightly archaic ring, asks whether competitiveness and aggression increase as land becomes scarcer and people more numerous. The answer from the writers of the present papers is clearly, "No". None of them reports a noticeable change in crime and unrest. Why not? Because, as these pressures mount, social forms change, enabling society to incorporate and support the increased population, thus rendering it no more asocial than the older generations. Societal organization and reorganization cushion the effects of population increase. These papers seek to describe the social rearrangements that occur when the ratio of man to land increases.

The pairing of Sri Lanka with Thailand for this study is not entirely fortuitous. Both countries face southern seas, raise common crops watered by the monsoons, have harbored Arab, Chinese and European merchantmen, once lived under governments demanding taxes and corvée labor, and today cherish Footprints, which according to tradition, attest to visits to these lands by the Lord Buddha during his life time. Through the centuries when India returned from Buddhism to Hinduism, Sri Lanka nourished the Theravada tradition as its center. Perhaps Thailand's great stupa at Nakhorn Pathom was a resultant. More certainly, in the 1300's Sukhothai kings petitioned for and received a delegation of Buddhist monks from Sri Lanka to rectify local practises and texts, thus strengthening the kingdom with merit and auspicious influences.

In turn a Kandyan king of the 1700's, uneasy with Dutch in the lowlands, turned to Ayuthia and welcomed a delegation of Thai monks to rectify local texts with a more perfect version. Nor does this exhaust the known and possible contacts between the two countries which, though rarely intimate, always remained aware of the other's presence.

To compare responses of agricultural villages and areas in these two similar countries is simpler than juxtaposing either, say, to Italy, Japan or China, with their industries, widely divergent climates and cultures. Because the significant rise of population in Sri Lanka and Thailand has occurred within the past century, much can be learned from accessible history. European commercial interests touched both countries in search of produce: Thai rice and teak; Sri Lankan copra, coffee, tea, rubber and spices.

Nevertheless, differences complicate the task. Sri Lanka, one eighth the size of Thailand in area and limited by the sea, has always been exposed to the world, while hills and jungles have insulated the continental Thai. The island's sharply defined topography still leads to strong contrasts between wet and dry zones, savannah and forest. There, human activities fragment into areas with special crops. In Thailand one section blends into the next, and few sections produce crops not shared with the others. Its single metropolitan front door on the sea has reduced contacts between foreigners and the people out back.

If we arrange these papers into a sequence on the basis of land availability and population density, a panorama of agricultural development seems to emerge. Enjoying the greatest resources in unused land and lowest concentration of population, first stand the Veddahs on the fringe of an unpopulated jungle with probably no greater population density than 25/km^2. They cultivate in shifting fields, though of late some have salvaged irrigation systems of long-dead villages for growing market crops. Next, the people of the Chi valley in northeastern Thailand cultivate in fixed fields, yet the land is still dotted with unkempt brush and patches of forest. Their scattered villages accumulate a population density of 140–200/km^2. In the central plains of Thailand Banoi's and its neighbor's single rice crop covers the entire area to the horizon, except for the more or less continuous strips along the canals given over to human habitation, where densities reach 200–300/km^2. (Atlas p. 44.) At the end of this continuum lies Mulgama in the Western Province of Sri Lanka, overflowing with plantations on the irregular lands, every tiny river valley terraced for rice and village sites constricted yet frequent enough to furnish a population density of 400–500/km^2. The impression of sequence is reenforced by the historical dimensions of the ethnographic papers which reveal such similarities as the concentration of population in a single, isolated village followed at some point in time by some inhabitants breaking away to form new villages on unused land. The contributions of Johnston and Peebles add convincing substance by showing changes in the land policies of Sri Lanka and the digging of canals in Thailand's central plains.

Let us turn to examining the data from the point of view of three assumptions about the nature of the man-land ratio. First, we shall assume that we are positivists dealing with an objective agent of social change. Second, we shall treat the man-land ratio as a phenomenal datum dependent on the particular interpretations given it by various social groups. Finally, we shall examine the ratio as a feature of social growth and decline.

Without necessarily foisting the thought on Selvadurai, let us presume that he observed Mulgama as a positivist and noted the following results: (1) multiple agricultural produce and two rice crops per year; (2) access to the soil limited by complex social devices for holding land; (3) a fine division of labor that distinguishes not only landholders from laborers but further sorts laborers into those with skills in tea, rubber, coconut and other types of production; (4) work recompensed in cash rather than by exchange; (5) a hierarchic valuing of social contribution giving prestige and command to landholders.

As we glance back over our course towards fewer people and more accessible land, we see these features of Mulgama drop away. On the Thai central plains they raise a single crop of rice rather than multiple crops. Banoi liberally allows land to be cultivated by any buyer or renter. Owners differ from laborers in prestige, yet both work side by side in the fields by day and eat from the same pot at night. Command falls to the landholder, but his authority does not outlast his ability to distribute wages and food. Much the same applies to the Chi valley of Thailand's northeastern plateau, though kenaf has been recently added to the limited production of rice. Labor may still be exchanged in many places, and free land for new settlements is not quite exhausted. Finally, for the unspecialized Veddahs, who supplement their agriculture with hunting and gathering, land not in use is freely available, and authority, based on age within the family, comes to all who can wait their day or to anyone who leads a few juniors away to form a new village.

Accepting this rough comparison, we glimpse the possibility of a correlation, or perhaps a causal relationship, between whatever the man-land ratio may represent in nature and the emergence of various societal characteristics as it increases. To establish such a correlation more firmly would require careful refinement of definitions for man and land as well as the emergent features of society. We would also require careful studies that will minimize or eliminate the contributions of other agents of social change such as the government-built canals and the influences of commercialization.

Whether or not this particular program would bear fruit, this positivist assumption underlies whatever study we phrase as to the effect of some agent on some other phenomenon. In this vein we might describe the present enquiry as the effect of rising man-land ratios on the social forms of agricultural villages. The agent, whether it be the man-land ratio, labor exploitation, or alcoholism, is presumed to be isolable and independent of the phenomenon on which it acts. I have no evidence that the initiators of this study considered or did not consider the man-land ratio in these terms.

Another group of inquirers with emphasis on the phenomenal as an agent of social change would find a gap in the evidence. For them social change occurs first in the mind of man and only later registers as actions and utterances of a people. From this point of view the man-land ratio contains a multitude of directives. To an adverse man-land ratio the fishing folk of Aran, that bleak, rock-bound island in the Atlantic off the western coast of Ireland, responded by gathering kelp for making soil. Chinese villagers of the past, when an adverse man-land ratio became acute, took to killing infant daughters and selling the older girls to some one in search of servants. In the Pacific islands an adverse man-land ratio resulted in launching a fleet in quest of another island. To such phenomenally-inclined students the ratio is scarcely a unitary agent, for it means land shortage to the people of Aran, population pressure to the Chinese and a time for embarking to the Polynesians.

Even among the present papers one can detect diversity of response. Though Keyes makes the case in the Chi valley for a perceived population pressure where people consciously make room by forming new villages at a distance, Brow suggests that the Veddah moved from their old to new villages to avoid conflict with newly arrived immigrants and with government officials hostile to shifting cultivation. In Piker's material Banoi coupled the man-land ratio with the desire to grow rich on selling rice, so pressed on, tilling up to the confining escarpment. Johnston reports feverishly restless people during the 1890's who moved to new land and then abandoned it the following year. Whatever moved them is not discussed, but clearly more than population pressure and land hunger governed their actions. Indeed, in the 1900's when Banoi began to divest itself of people, we learn, as Piker quotes Zimmerman, that the average area for cultivation was 28 rai or 11 acres of land per household. When we compare this figure with Mulgama's less than an acre of paddy land per household, an objective man-land ratio fades into insignificance as an agent of social change.

But these various perceptions too cannot serve as sufficient agents of social change, for they act in concert with many other perceptions. Population pressure and land hunger arose in Thailand only after the cultivators saw waterways leading into the eastern wilderness, so that we may as well attribute the changes to the Royal Thai government, the Siam Land, Canals and Irrigation Company, its workers or the host of other entrepreneurs. Peebles adds that the highlands of the Western Province in Sri Lanka were a jungle until the Colonial government proclaimed land to be available in fee simple, and markets encouraged coffee or tea production. Until then the Sinhalese saw no value in these lands. Thus phenomenally-oriented students return with data in tangled webs, no strand of which can be tugged even gently without further entangling the whole mass. To follow the strands calls for a hero greater than Alexander who can untie rather than cut the Gordian knot.

It would seem that the man-land ratio clothed as phenomenon is hopelessly enmeshed in all kinds of change. One further guise remains to be tried, that the man-land ratio be treated as a symptom, the significance of which can only be assessed when social change is given in a context. The initiators of our collaboration, however, did not happen to provide us with such a context, and aside from occasional lamentation about a division of labor that effaced a village-wide equality of status, the direction of social change is left for the reader to infer. Yet social change, even "structural" change, is ubiquitous, so that until it receives a context, we are merely viewing a kaleidscope.

In seeking a context for social change, we encounter again the horrors that began our peregrinations, the agonized destruction of a social group. Social death and survival, despite their apparent clarity, pose formidable problems of definition. For instance, if Poland survived partition in 1795 and still lives in the Polish People's Republic, does Bohemia continue to live in Czechoslovakia? Evolution also furnishes a context by building stages and sequences of change, yet since many stages and sequences can be bypassed, the readily available formulations are inadequate for our purposes. More modestly, let us posit growth and decline to frame social change. A social group grows in population, productivity, and the elaborateness of its social institutions, declines when these indices decrease, or sustains itself when they remain fairly constant. Such a scale is rough but adequate for a preliminary framing of social change.

In this context the papers speak more strongly, for we are witnessing the growth of social groups both in Sri Lanka and Thailand. All these villages and areas have actively utilized population to expand production, have increased lands available, and modified their institutions to manage land and people more effectively. In the course of growth, problems emerge which limit or threaten to limit growth. One such problem is the adequacy of land, another the adequacy of water, a third, the integrity of the social organization when command becomes diffuse. The Veddah and the Thai solved the land problems by moving people to new sites. The Sinhalese of Mulgama not only extended their rice-producing land but diversified their crops to unused land, as did the people of the Chi valley. Where water was short the Veddah rebuilt old irrigation systems, while the Thai of the central plains pumped canal water into their seed beds. By keeping villages relatively small, the command problem was bypassed, but other problems still face these villagers as Fraser's observations of the same Chi valley show:

> "Wherever adequate roads exist, trucks and buses have replaced bullock carts for the transport of produce to the market. Many farmers engaged in producing garden crops have found that renting small gasoline-powered irrigation pumps during the dry season will significantly increase their yields and profits. Of perhaps greater importance is the rather rapid adoption in areas near urban centers of renting tractors from merchants in town to plow the rice fields before the advent of rains. Prior to this innovation the farmer had to wait until the rains had softened the ground before his buffalo could draw the plow through it." (Fraser)

When commerce enters, transport and the increased production pose additional problems. The Thai are solving them with machines that are less available to Sri Lankans, who cope just the same.

In contrast to this scene we may note Ireland with population declining, a shrinking variety of produce, a reduction of parishes, schools and other social arrangements, an ebbing that began with the Potato Famine. Among the American Indians, populations have increased after a sudden drop in the late 1800's, but on most reservations dispirited people have neither increased production nor elaborated their institutions. Here we may sense that one ingredient of growth is morale, which in its upper reaches heartens people, still weary and black from a conflagration, to rebuild rather than abandon their homes. Certainly these papers are also saying that population pressure and land shortages are likely concomittents of social growth but have not yet imposed overt limits of growth on the cases within our purview.

Limiters of social growth do exist, but only one symptom was described that might carry this potential. In Mulgama access to land contracted to the point that some village members appear to have been excluded from raising crops. The text, however, omits the sequel: whether the affected households departed or enriched the village with new occupations. In India the limitation of social growth requires no elaboration. Myrdal diagnoses it as the product of "underutilization of labor", thus suggesting a limitation of the social mechanisms that allocate tasks and motivate people (Myrdal 1968: 1356). Even though Geertz depicts in Java changes almost identical to the ones described in these papers, he finds a limiting factor in an "advance toward vagueness". By this he indicated a kind of spiritless resignation that arose from decades of governmentally imposed restraints on Javanese agricultural expansion, which became all the more evident in contrast with the government's favors accorded Dutch commercial planters (Geertz 1963: 102–103).

We are left with uncertainties. May we conclude from these papers that the man-land ratio poses a problem in social growth but offers no greater limits than many other problems of a growing group? Or is the man-land ratio in some respect more inexorable or baffling than, say, marginal rainfall, soil depletion, or malnourishment? If not, should we conclude that the spectre of population increase and land shortage has been exorcised? Is it more likely, perhaps, that this spectre has just been driven into a remoter corner? The style of its threat, in any case, seems to have changed the horrifying end from howling violence to apathetic stillness.

REFERENCES

1964 *Atlas of Southeast Asia*. MacMillan, London.

Fraser, Thomas M., Jr.
1974 "Modernization of Peasant Society in Northeastern Thailand". *Annals of the New York Academy of Sciences*, Vol. 220, pp. 376–384.

Geertz, Clifford
1963 *Agricultural Involution*. University of California Press.

Myrdahl, Gunnar
1968 *Asian Drama*. Pantheon, New York.

Contributions to Asian Studies, Vol. IX

The Closing of the Frontier

Land Pressures and their Implications for Rural Social Organization in the Thai Central Plain

STEVEN PIKER

Swarthmore College, Swarthmore, U.S.A.

> *(Sisuriyawong)* Now... Sir John Bowring: answer me: why do you come to Siam?
> *(Bowring)* We are to become your neighbors—we wish to be your friends... Don't shut the door...
> *(Sisuriyawong)* Well, in a large house with many servants, the door may safely be left open; in a small house with few servants, the door must be closed... (and, later on) yes, a soil, but no people. A soil without people is but a wilderness...[1]

IN THIS VEIN did *Somdet Chao Phraya* Sisuriyawong address Sir John Bowring, the British plenipotentiary who in 1855 negotiated with King Mongkut the treaty which opened Thailand to European economic penetration and set in train a series of events which, during the century that followed, transformed all levels of Thai society.[2] Much land, but few people, was a chronic condition of the Thai state into the early years of the 20th century, and a circumstance with far reaching cultural and institutional implications. We may begin by tracing the historical dimensions of this circumstance—or problem, as the rulers of the Kingdom perennially found it to be—and go on to consideration of very recent demographic changes, and their consequences for rural social organization in the Central Plain.

1. Land, population, and society in the Ayuthayan period (1350–1767)

Demographic patterns for Thailand before the mid-19th century are at best poorly known. Ethnic Thai probably began filtering into South-east Asia in significant numbers in the 10th or 11th centuries, A.D., the pattern almost certainly being uncoordinated migration and not invasion.[3] Much of the area which today comprises Central, North, and Northeast Thailand was from the

---

1 Bowring, Sir John, *The Kingdom and People of Siam*. p. 464ff.

2 Prior to the signing of the Bowring Treaty, relations between Siam and England were governed by the Burney Treaty (1826). The Burney Treaty left Siam the right rigorously to restrict trade within its borders, and contained no provisions for extraterritoriality.

3 See, for example, G. Coedes, *The Indianized States of South-east Asia*. p. 189ff.

11th to the 13th centuries under the suzerainty of the Khmer Empire centered upon Angkor in the heart of modern day Cambodia, and in all three regions the numerically dominant population was almost certainly Mon-Khmer. Just as certainly, the population of these regions was sparse, being largely restricted to a handful of major towns or centers and to rural settlement along major waterways and trade routes. As the numbers of ethnic Thai in SE Asia increased, a major demographic rearrangement began, which in its early stages elevated the newly arrived Thai from minority to coordinate status, and which subsequently provided them with clear numerical ascendancy—in much of Laos and NE Burma as well as those regions (excepting only the South) which today comprise the Thai Kingdom.

By the close of the 13th century, many of the Thai ruled principalities of Northern Thailand had been united under the dynamic leadership of Rama Kamhaeng's Sukhothai, and since that time centralized Thai governments have ruled almost continually over most if not all of the area that comprises modern day Thailand as well as, periodically, more remote regions which today are incorporated in one or another of Thailand's neighbors. For most of the Ayuthayan period ethnic Thai were numerically dominant in the Central and Northern regions, in part as the result of natural increase and immigration, and in part as a result of assimilation of Mon-Khmer population. During the Bangkok period (1782–present) ethnic minorities—excepting only the mountain tribal peoples of the North and the mainly urban Chinese—in the Central and Northern regions have become numerically inconsequential.

The increasing political and demographic ascendancy of ethnic Thai throughout most of the regions which today comprise the Kingdom did not, however, bring with it an increase in population density. Far from it. Until quite recently, the land area controlled by the Kingdom was immense by comparison to the population which inhabited it; and within the Kingdom vast stretches of land were either sparsely settled or uninhabited. No reliable population estimates for the Ayuthayan period are available. For the mid-19th century, however—and it is doubtful that the population of the Kingdom had ever been greater—a number of authorities[4] estimate a figure of roughly five million, of whom a substantial number were inhabitants of remote Laotian and Malayan dependencies. Taking this into account, the combined population of the Center and the North in the mid-19th century probably did not exceed 2.5 million, which is less than 15% of the population for these two areas in 1970.[5] For the larger part of the Ayuthayan period, the population for these two regions must have been lower than this figure.

An immediate corollary was that all Thai rulers—from the early Ayuthayan Kings to the Chakri Monarchs of the late 19th century—were confronted with the problem of retaining control of a domain which could not

---

4 Bowring, *op. cit.*; John Crawfurd, *Journal of an Embassy to the Courts of Siam and Cochin-China*; Mgr. Pallegoix, *Description du Royaume Thai ou Siam.*

5 Thomlinson, Ralph, *Thailand's Population: Facts, Trends, Problems, and Policies.*

be fully or evenly populated with loyal subjects. Thailand, in short, for most of its national history, has been a frontier society, and its emergent legal and social institutions have been importantly influenced by this circumstance. Here we touch upon two aspects of the picture, both intimately related to the population problems of the Kingdom.

Slavery

The origins of bondage in Thailand are obscure. Sukhothai inscriptions, however, testify to its existence late in the 13th century, and doubtless the practice continued uninterrupted until the end of the 19th century.[6] Two points as regards slavery are particularly important for present purposes. First, the practice almost certainly originated in response to the manpower needs of the Kingdom. Not just prisoners of war, but usually whole villages or even districts from the defeated country were transported by the victor to his native land, settled there, and subsequently assimilated to the host ethnic group. The point, clearly, was simultaneously to increase Thailand's population and to decrease that of neighboring kingdoms, whose strength, just as Thailand's, depended ultimately on numbers of subjects. This practice continued in to the first half of the 19th century, when Lao villages were removed to Central Thailand following a resounding Thai victory in Laos. We visited a cluster of such villages near Lopburi town in 1962; they are now indistinguishable in appearance from surrounding villages, but their inhabitants know that their ancestors were brought down from Laos as war booty in the 19th century.

Second, although slavery originated in response to the population needs of the Kingdom, the institution as it developed came to have far wider significance, principally for native Thai. War captives came in fact to comprise by far the smaller portion of the slave (*that*) population of the Kingdom as the institution came to take on aspects of debt bondage. Many rural Thai entered bondage voluntarily, often to discharge debts but sometimes simply in preference to the rigors of life as a freeman; nor was it uncommon to sell at least some of one's children into bondage to discharge debts. The practice was legally regulated to insure that numerous rights and protections accrued to slaves; and in practice the slave system in Thailand probably resembled more closely a patron-client kind of arrangement than slavery as it has been known in the West, particularly in the United States.[7] Although its numerical extent is not reliably known, the number of slaves for the early 19th century has been estimated as high as one fourth of the population of the Kingdom.[8] This probably errs on the high side, but bondage was unquestionably a prominent fixture of Thai rural life until its abolition during the reign of King Chulalongkorn at the close of the 19th century.

---

6 Wales, H. G. Q., *Ancient Siamese Government and Administration*, p. 58.
7 *Ibid.*, p. 63.
8 Bowring, *op. cit.*; G. Lasker, *Human Bondage in SE Asia.*

Access to unused land

Settlement patterns in Central Thailand until the beginning of the 20th century had most of the population in the capital city (first Ayuthaya, subsequently Bangkok-Thonburi), a number of towns, and—in rural areas—strung along the banks of major water ways, namely, the principals rivers and those canals that had been established. The low population levels of the Kingdom would have precluded settling most or all of the region in any event; but, this aside, peasant cultivators depended upon the waterways for communication (there were few roads in the countryside) and, especially, for the yearly flood required by wet rice agriculture. The immense hinterland—seemingly endless tracts of forest and marsh—began nowhere more than a few miles from areas of even densest settlement; and although Thai Kings may have laid claim to domains stretching for hundreds of miles, their effective power stopped at the limits of settlement (if indeed it always stretched that far). Beyond lay unpopulated and unwatched regions, in which predatory outlaws could exist in comparative safety and through which enemy armies might approach unnoticed and uncontested until they arrived before the gates of the capital or a major town.

Given these conditions, it was very much in the interests of the monarchy to encourage the extension of settled and cultivated areas, especially in the Central regions adjoining the capital, insofar as population permitted. Consequently, the legal structure of the Ayuthayan state as it evolved gave conspicuous attention to regulation of population, and especially to the relationship of population to land. One aspect of the emerging picture—those provisions intended to insure that the services, either military or corvee, of the peasant freemen would be available to the state and its functionaries—requires no treatment here.[9]

Another, however, having to do with the legal provisions by which access to unused land was regulated, must be mentioned.

The lay population of Ayuthayan society was organized hierarchically with the monarchy at the apex.[10] Below the monarchy the fundamental social distinction was between *nai*—lords of the nobility who, as functionaries of the King, were responsible for managing the population and resources within their purviews—and *phrai* (free commonors) and *that* (slaves). Commonors and slaves were directly accountable to lords, who at least in theory were accountable to the King. The yardstick of rank was provided by the *sakdina*, or "dignity marks" system. The etymology of the term *sakdina*, which to Wales[11] suggests "power of (or over) fields," is instructive. As one authority[12] notes, "There appears to have been a strong correlation between the amount of *sakdina* one

9 For a detailed discussion of this issue, see: Wales, *op. cit.*; Rabibhadana, Akin, *The Organization of Thai Society in the Early Bangkok Period*, 1782–1873.

10 Wales, *op. cit.*; Rabibhadana, *op. cit.*

11 Wales, *op. cit.*

12 Abibhadana, p. 22.

possessed and the number of people one had beneath one's control... originally *sakdina* was concerned with the amount of land one was allowed to possess... The reason for the limitation was to allow people to have only that amount of land they were capable of developing... therefore... the amount of *sakdina* and the extent of land possessed by any one person grew with the increase of the number of people under his control." For free commonors, increase in *sakdina* could be achieved only by founding a family (just as the King, a *Chao Chiwit*, was lord—or owner—of all life in his domain, so also did the husband/father legally "own" other members of his family; he could, for example, on his own authority sell some or all of them into bondage); and access to land increased proportionately. The *sakdina* of an unmarried free commoner was 15, but could increase to as much as 25 as his family grew larger; and, accordingly, free commoners could legally control up to 25 *rai* of land (1 rai = .4 acres).[13]

Theoretically all of the land in the Kingdom was owned by the King, and the rights of free commoners were restricted to usufruct. In practice, however, use and de facto ownership merged. Land was transmitted from generation to generation in the families of freemen; indeed, kinship institutions and the domestic cycle, as they evolved as the mainstays or rural social organization, revolved most concretely around inheritance practices[14]. Land also could in effect be sold by freemen much as private property. The legal provisions of the *sakdina* system, then, along with their informal extensions, encouraged free commoners to avail themselves of unused land when they had the necessity and the resources to do so, providing that they maintained formal ties of accountability and subservience to a lord. Even this last stipulation, however, did not in fact much inhibit the mobility of freemen into unused lands; for along with it went the provision that under various circumstances commoners could shift their allegiance from one lord to another. Since the power of lords, officially, according to the *sakdina* system depended largely on the number of commoners under their control, they were unlikely to give commoners cause to defect by placing too tight a hold on them; and doubly so since they knew that other lords would be happy to enhance their own power and prestige by providing a legal haven for disaffected commoners. For these reasons, an essentially feudal system which had developed in part for the purpose of establishing tight control over population also in effect encouraged a degree of geographical mobility on the part of the population, as well as a degree of flexibility—unusual for feudal societies—in the relationship between lord and commoner. The double administrative exigency with which the Ayuthayan

13 It is noteworthy in this connection that even slaves had a *sakdina* of 5. It could not increase —for the slaves' children as well as the slave belonged to his master; nor could a slave own land. Still, the fact that the slave had a rung on the ladder, albeit the lowest one, symbolizes his inclusion within the broader society and underlines the legally defined rights and protections that were extended to him.

14 For an extended discussion of village social organization, including kinship institutions, see: Piker, "Peasants in a post-peasant society."

Kingdom was confronted—control of population, and control of territory—in effect dictated this compromise, a consequence of which was that disinherited or otherwise restless freemen could turn to the frontier to begin a new life.

To summarize: a number of the dominant features of rural Thai society in the Central region as it evolved during the Ayuthayan period were shaped by the following factors: very low population density; a legal system which made unused land easily available to freemen; and bondage as a viable and not unattractive alternative to life as a freeman if, for whatever reason, its rigors became insupportable.

II. The 19th century

Ayathaya was captured and sacked by invading Burmese armies in 1767, and internal chaos followed. As had happened following similar disasters in the past, however, the strength of the Thai nation quickly reasserted itself. By the end of the 18th century a new Thai dynasty—the Chakri—was in control of most of the regions ruled by the Ayuthayan Kingdom in the mid-18th century; the invading Burmese armies had been defeated and expelled; internal order had been largely restored; and the capital had been removed fifty miles to the south, first to Thonburi and shortly thereafter across the Chao Phraya River to the site of modern day Bangkok.[15] The primary assumed task of the first three rulers of the new dynasty (Rama I-III, 1782–1851) was to restore the power and institutional fabric of traditional Thai society, albeit with specific modifications, not to remake the society; and in these goals they were notably successful. The mid-19th century, however, brought a new and wholly unprecedented challenge to Thai sovereignty, with the advent of European imperialism. And it was largely in the interests of maintaining national sovereignty that King Mongkut (Rama IV, 1851–68) and King Chulalongkorn (Rama V, 1868–1910) after him launched the internal reforms that were to undo once and for all the Ayuthayan institutional synthesis and set the direction for social change in Thailand for a century and more. The details of this process—its initiation and continuation, often in directions unforeseen by its progenitors—cannot be treated here. Since, however, the Bowring Treaty signified so importantly both the initiation and the direction of subsequent changes in Thai society, it will concern us briefly. Although the institutional consequences of the Bowring Treaty and the policies it brought in its train ramified throughout all sectors and strata of the Kingdom, we are here concerned mainly with its implications for the organization of rural society in the Central Plain.

I have elsewhere summarized some of the significance of the Bowring Treaty as follows:[16] "Most importantly, Thailand's longstanding ban on rice exports

15 Wenk, Klaus, *The Restoration of Thailand under Rama I, 1782–1809.*
16 "The post-peasant village in Central Plain Thai society."

was abolished, import duties were set for most commodities at the negligible level of 3 percent (opium was to be imported free of duty), export taxes were both simplified and drastically reduced, and British subjects were granted the right of free trade in all ports. (Between 1856 and 1899, fourteen other countries —all Western with the exception of Japan—concluded treaties with Siam on the model of the Bowring Treaty.) The immediate and anticipated consequence of these arrangements was the articulation of Thai economic institutions— previously (almost) entirely insular—with international markets. Derivatively, the social and economic organization of the Kingdom was fundamentally remade. Thailand's increasing participation in international markets... placed a premium on expansion of rice production and extensification of cultivation. It was recognized that the labor increment required to achieve these goals could be realized only by the relaxation of the obligations entailed by (the universal service obligation of the peasantry to the state) as well as (abolition of slavery)... Since the revenues of the Kingdom were expected to increase dramatically as a result of burgeoning trade, it was anticipated that the state's labor requirements—previously met largely through corvee levies—could soon be met by hired labor, and that slave owners might be able to find monetary compensation for the gradual abolition of bondage. Accordingly, slavery was gradually but systematically abolished during the last decades of the 19th century, and the corvee labor system was transformed into an arrangement in which categorical obligation (was supplanted by) wage payment—a development no doubt facilitated by the rapid influx during this period of impoverished Chinese, many of whom welcomed this opportunity for employment.

The implications of these developments for rural society in the Central Plain were far reaching. A peasantry, both legally and actually free, had now to adjust to an agricultural economy oriented increasingly to production for markets as opposed to local consumption. Moreover, as a consequence of extensification of cultivation in response to favorable market conditions and a dramatic increase in population... the availability of unused land—historically, and into the first years of the 20th century, one of the bulwarks of a relatively stable Central Plain rural society—came to an end."

The magnitude of the transformations which came in the wake of these developments is indicated by the following tables and summaries.

If the 1.5–2 million estimate for Central Plain population in 1850 is accepted, it is seen that the population of the region approximately doubled between 1850 and 1911, and more than doubled again by the end of World War II. Of the absolute increase, from between one and a half and two million (estimated, 1850) to seven million (1947), less than 10% can be attributed to the growth in Bangkok's population, the main area of settlement for newly entering Chinese; so it may be supposed that by far the larger part of the growth was the result of increase in the rural population, which was almost entirely Thai.

Of even greater significance than population growth, however, in its im-implications for rural social organization, was the extensification of cultivation

Table I

Thailand's population, 1850–1960[17]

Date	*Estimate*	*Population ('000)*	*Average annual rate of population growth since previous estimate (%)*
1850	Ingram*	5,500	—
1911	census	8,266	0.7
1919	census	9,207	1.4
1929	census	11,506	2.2
1937	census	14,464	3.0
1947	census	17,443	1.9
(1956)	survey†	(20,095)	—
1960	census	26,258	3.2‡

* Ingram (1955: 7–8).
† National Statistical Office (1957b: 1). Note that the estimate is not for 1954 but for 23 February 1956.
‡ Calculated for the period 1947–60.

Table II

Thailand's population, 1911–1947, the Center compared to the remainder of the Kingdom[18]
(000 Omitted)

Date	*Center*		*All Other*		*Total*	
	Population	*Index*	*Population*	*Index*	*Population*	*Index*
1911	3,267	100	4,999	100	8,266	100
1919	3,520	108	5,687	114	9,207	111
1929	4,582	140	6,924	139	11,506	139
1937	5,748	176	8,716	174	14,464	175
1947	7,000	214	10,317	206	17,317	209

Sources: *Statistical Year Book of Thailand.*

that occurred during roughly the same period in reponse to a thriving export market.

The extensification of rice cultivation in the Central Plain beginning in the last half of the 19th century has, as inspection of table 4 shows, been considerable. Market demand generated by the new rice exporting policies provided the impetus; and the peasantry of the Central Plain, freed from traditional labor obligations to lords and state, responded to income incentives by

17 Reproduced from: Caldwell, J. C. "The demographic structure." in T. H. Silcock (ed.), *Thailand: Social and Economic Studies in Development.*
18 Reproduced from: Ingram, J. C., *Economic Change in Thailand, 1850–1970.*

Table III
Increase in rice exports, 1857–1951[19]

Period	*Average Volume per Year (Thousand Piculs)*	*Average Value per Year (Thousand Baht)*	*Average Price per Picul (Baht)*
1857–59	990		
1860–64	1,840		
1865–69	1,630		
1870–74	1,870	5,110	2.70
1875–79	3,530	10,110	2.90
1880–84	3,580	9,610	2.70
1885–89	5,320	15,080	2.80
1890–94	7,250	23,780	3.30
1895–99	8,000	36,410	4.60
1900–1904	11,130	61,280	5.50
1905–09	14,760	81,020	5.50
1910–14	15,220	81,230	5.30
1915–19	15,790	108,140	6.90
1920–24	17,680	115,350	6.50
1925–29	23,390	169,600	7.20
1930–34	25,720	91,240	3.50
1935–39	25,370	94,570	3.70
1940–44	13,250	99,320	7.50
1946	7,580	267,340	35.30
1947	6,535	384,605	59.00
1948	13,540	1,255,335	93.00
1949	20,260	1,869,410	92.00
1950	24,820	1,996,190	80.00
1951	26,290	2,223,610	85.00

Table IV
The growth of area planted in paddy, 1850–1950[20]

Period	*Center*	*All Other*	*Total*
1850			5.8[a]
1905–9	6.8	2.4	9.2
1910–14	7.2	4.3	11.5
1915–19	8.2	5.7	13.9
1920–24	9.6	6.7	16.3
1925–29	0.19	7.2	18.1
1930–34	11.8	8.3	20.1
1935–39	12.0	9.2	21.2
1940–44	14.1	11.4	25.5
1945	12.1	11.4	23.5
1948	13.1	11.8	24.9
1947	14.6	15.6	30.2
1948	15.8	16.8	32.6
1949	16.3	16.6	32.9
1950	16.7	17.9	34.6

(Unfortunately, no estimate for the Center for 1850 is available. If it is recalled, however, that the total land area of the Kingdom at that time, including as it did virtually all of the area of modern Laos and Cambodia, was almost twice the present area, it seems reasonable to suppose that the area under cultivation in the Center could not have been more than half of the total land cultivated in the Kingdom, and in fact was probably considerably less than half.)

bringing new lands in all directions under cultivation. Population increase (tables 1 and 2) doubtless facilitated the process. So also did a rise in the land to labor ratio,[21] signifying that villagers were prepared to invest more hours in rice cultivation—the labor requirements of which are in any event seasonal—if prospects for increased income were good.

The extensification process did not occur evenly throughout the Central Plain, however. Understandably it proceeded most rapidly where access to waterways—which in most areas provided the only means by which paddy could be transported to mills and thence to markets—was easiest. For this reason, commercialization with attendant extensification of cultivation occurred earliest in the middle regions of the Central Plain, in and near Ayuthaya Province, where major canals already existed and could be extended with relative facility. Only later, in the closing years of the 19th century and into the 20th, under the impetus of government sponsored canal building programs, were the southern expanses of the Central Plain opened up. Paradoxically, some of the last regions in the Central Plain to be opened up by canals, settled, and brought under yearly cultivation were only a few leagues removed from Bangkok, and in these areas the process was not completed until well into the 20th century.[22]

Notwithstanding unevenness of rate, the outcome was remarkably uniform throughout most of the Central Plain: by the quarter mark of the 20th century, most previously unused land in the Central Plain had been cleared and brought under cultivation; production in most areas was by that time predominantly for market and not local consumption; and the day was past when uncleared land abounded within a few miles of most densely settled areas, available to anyone who would face the rigors of domesticating it.

III. The village of Banoi in the 20th century

The village of Banoi has been in the direct path of the events just alluded to. Located five miles (8 kilometers) due east of the town of Ayuthaya, Banoi and its environs have been the site of established Thai villages certainly for three centuries, and quite possibly for considerably longer.[23] Since the early Ayuthayan period, this region of the Central Plain has been the granary of the Kingdom and, for this reason, major canals have long been established to supplement the major rivers for communication and transportation purposes. When, following the signing of the Bowring Treaty, national policy called for rice surpluses to export, the environs of the town of Ayuthaya were among the

---

19 *Ibid.*, p. 38.
20 *Ibid.*, p. 44.
21 *Ibid.*, p. 55.
22 For an extended description of the opening up of regions close to Bangkok, see: L. M. Hanks, *Rice and Man: Agricultural Ecology in SE Asia.*

first to respond. These were regions of already proven high yields, and water links already existed to carry the milled rice to the port of Bangkok.

If the forces of commercialization and extensification of cultivation were felt early on in Ayuthaya Province, however, it should nonetheless be emphasized that these were quite recent developments. Although no villager living in or around Banoi in the 1960s could remember the reign of King Mongkut, a number of the older ones did remember the situation around the turn of the century, and the content of their memories is instructive. At that time, the Eastern limit of continuous settlement in Ayuthaya Province was only a few miles from Banoi. Beyond this ill-defined line one found an occasional cluster of houses near patches of cultivated land, but no established villages, much less towns with markets and administrative personnel. Here the jungle and the marsh ruled for the most part, all the way to the range of low mountains—some thirty to forty miles further east—which separate the Central Plain from the Khorat Plateau. Villagers in their sixties and seventies report that during their childhood tigers were hunted only a few miles away; and wild elephants were occasionally rounded up a day's or two days' march to the East—possibly no more than ten miles from the village. Nor did the rule of the state, such as it was, extend beyond the limits of established villages. Brigandage was much more common around the turn of the century than it since has become, doubtless in part because bands of thieves could flee at least temporarily to the wilds when government soldiers came in pursuit of them.[24]

These conditions have vanished in the lifetimes of our elderly villagers, and today the flat lands to the east of Banoi are cleared, settled, and cultivated all the way to the mountain barrier. Interestingly, however, local vernacular continues to reflect the frontier situation that existed in Ayuthaya Province into the first decades of the 20th century. The region beginning about three to four miles east of Banoi, beyond the *amphur* center, is still universally referred to be local inhabitants as the *paa*, or forest, which term for them carries definite connotations of both geographical and social marginality. References to this region—the *paa*—as a den of thieves and a source of unusual danger to visitors are commonplace, although no longer grounded in fact; and strangers passing through Banoi who are known or believed to have come from the *paa* are viewed with considerably more suspicion than are strangers from elsewhere. For older villagers, circumstances which actually obtained in the *paa* during their youth justified these perceptions, even if contemporary conditions no longer do.

The commercialization of rice agriculture in the heart of the Central Plain during the decades surrounding the turn of the century has been dealt with

23 Other Central Plain villages described in the anthropological literature on Thailand were founded as a result of the late 19th century developments described in these pages. See, for example, Hanks, *ibid.*, and H. K. Kaufman, *Bangkhuad: A Community Study in Thailand.*

24 It was also true that, earlier in the 20th century, villagers often were prepared to shield brigands from the forces of the state—sometimes under threat, but sometimes because of a fundamental sympathy with either the person or the ostensible purposes of the brigand.

in detail by Ingram,[25] and requires no further discussion here. With the vanishing of the *paa*, however, there came three fundamental changes in the circumstances of villagers of Banoi which must command our attention. First, general population increase along with the disappearance of unused lands available to villagers prepared to take advantage of them produced substantial local population growth.[26] The memories of villagers are unanimous in this regard. Population grew steadily, and new houses and compounds were erected, throughout the first decades of the 20th century, until a peak was reached during the 1930s. Concerning conditions during this period, an Abbott of a local *wat* who was then a young man remarked, "If a bird had flown over the village, he could not have found a place to land, it was so crowded." A period of gradual out-migration followed, in response both to growing landlessness and employment opportunities elsewhere; it continued into the 1960s, producing an estimated 25% decline in local population.

Second, as a result of developments described above, landlessness on an unprecedented scale developed in the decades prior to World War II. Zimmerman[27] reports a landlessness rate for Ayuthaya Province in 1930 of 42% (and an average of 36% for eleven sample Central Plain provinces), which signifies an advanced stage in the development of this pattern even at this early date. There can be little question that landlessness in socially significant amounts was present in Ayuthaya Province throughout the early decades of the 20th century. Family histories collected in the Banoi region in the 1960s show that this was so. For Banoi and neighboring hamlets in the 1960s, the rate of landlessness approached 50%.

Third, if commercialization of agriculture brought pressures that produced landlessness for some, it also brought opportunities that made possible unprecedented prosperity for others. Successful rice cultivation was now a money making business. Whereas previously a bumper crop meant unusual short run security and plenty for one's family along with the ability to assume, if only temporarily, the status of informal patron vis-a-vis less fortunate kinsmen and neighbors, it had now come to mean possession of investable capital as well. And the object of investment for newly profit conscious cultivators was land upon which more salable rice could be grown. The result over the years was an increase in the size of many landholdings as successful cultivators liquidated the debts of their less fortunate neighbors by purchasing their paddy fields from them when sale could no longer be avoided. Landholding figures for the Banoi region for the pre-World War II period are not available; Zimmerman, however, shows an average of 26 *rai* owned by landed families in Ayuthaya Province for 1930.[28] If in-rentals are taken into account, the amount of land

25 Ingram, *op. cit.*

26 The growth of population in Ayuthaya Province during this period is attested to by the establishment of a number of new district, or *amphur*, centers within the province for the purposes of administration.

27 Zimmerman, C. C., *Siam: Rural Economic Survey, 1930–31.*

28 *Ibid.*

farmed by many families, especially successful ones, would have been five to ten *rai* more than this figure. Estates of this magnitude were quite large by traditional standards. In other regions of the Kingdom in 1930, where commercialized agriculture was yet in its infancy or unknown, average holdings were often no more than a third or a half of this figure. Indeed, the reader will recall that the *sakdina* system effectively prevented peasant cultivators from owning more than 25 *rai* of land, and that the major reason for this legal restriction on size of holdings was that one family could not on its own hope to cultivate more than this amount. It follows that many landowners in Ayuthaya Province by 1930 farmed considerably more land than could be cultivated by the labor of a single family, even if their efforts were supplemented by the kind of assistance, given on a reciprocal basis, that was and remains commonplace among closely related families. This conclusion has special force for the minority of families which, by 1930, were farming more than 50 *rai* on a yearly basis. (It may be noted parenthetically that the technology of rice agriculture in 1930 was very much what it had been a hundred years earlier; there is no reason to suppose that commercialization led to greater efficiency in production.)

These three developments heralded a major change in the social organization of agriculture. Whereas traditionally, and into the first decades of the 20th century, peasant freemen had cultivated their plots with the labor provided by their families supplemented by that received on a reciprocity basis from close kinsmen or, sometimes, unrelated neighbors, by the 1930s it had become necessary for increasing numbers of them to hire wage laborers as well. That they were able to do so was the result in part of their own new found prosperity, an important aspect of which was the yearly availability of unprecedented amounts of cash deriving from sale of paddy. It was also due to the fact that the same circumstances which had made possible their new found success in commercial agriculture were producing a growing quota of landless villagers who, if they wished to reside in the locale of their birth, perforce turned to wage work as the basis for subsistence.

By the 1960s these trends had reached an advanced stage in and around Banoi. Farming families cultivated, on the average, almost 50 *rai* of land, including in-rentals, and all such families made extensive use of agricultural wage laborers in their farming enterprises. By the same token, landless families in Banoi and neighboring hamlets had come to outnumber their landed counterparts in the ratio of 5:4, and subsistence for most landless families depended largely upon wage labor performed by at least some of their members on the estates of their landed neighbors.

The following brief case histories illustrate these developments, and point as well to central features of a newly emerging social pattern in the countryside. 1) Daeng and Nawm were married in the 1930s. Both came from families that were well to do by local standards, and they received more than 30 *rai* of rice land at the time of their marriage from their respective parents. Although the common practice is for the newly wed couple to receive land mainly or only from the bride's family, which itself receives bride price from the groom's family in return for his actual or residual rights in their land, in this case about equal amounts of land came from each family. As is the local custom, they resided matrilocally following marriage, in the compound of Daeng's parents. When Daeng's two younger sisters married, they also brought their husbands into their parents' compound and family, the youngest living with her husband in her parents' house. The three married couples, children when they appeared, and the sisters' parents comprised a working kindred. Although the families owned their lands separately, they exchanged considerable assistance—both labor and buffaloes—during the farming seasons. In the early years of their marriage, the labor of Daeng and Nawm, supplemented by the help they received on a reciprocity basis from their kindred co-members and the services of a client family (to be discussed below), approximately met the needs of their farming enterprise. If especially for harvest, which of all stages of the agricultural cycle requires the most labor, further assistance was needed, they had recourse to a quasi-cooperative labor donating arrangement called *longkaek*.

Due mainly to diligence, hard work, and probity on the part of both husband and wife, the fortunes of Daeng and Nawm prospered. When we first met them in 1962, their landholding had increased to almost 80 *rai*. They still resided in the same compound, with Daeng's youngest sister and her now elderly and retired mother living in an adjacent house (Daeng's father had died); and they still exchanged labor and other services regularly with Daeng's sister's family, as well as with a family more distantly related to Nawm which resided in his natal village a couple of miles away. In all of these respects, their agricultural enterprise was run more or less in the traditional manner. If, however, this labor input had been sufficient to farm 30 *rai*, it was woefully inadequate for more than twice this amount. To make up the substantial labor deficit, Daeng and Nawm had for many years been hiring agricultural wage laborers in large numbers. These fell into two categories: first, migrant laborers who resided non-locally and who travelled through the Central Plain at harvest time seeking wage work (and, beginning in the 1960s, roving tractors owners who rented their tractors and driving services for the first plowing); and, second, landless neighbors whose yearly incomes came largely from the wage work they could obtain locally during the planting and harvest seasons. Daeng and Nawm established relationships with the local laborers that were open, informal, and diffuse compared to the relationships they had with migrant workers. The former were, after all, neighbors and sometimes casual friends (almost never were they kinsmen). Nevertheless, the arrangement with both categories of laborers was invariably contractual, revolving around payment of an agreed upon wage for a fixed amount of work; and the employee-employer relationship, even if the parties were neighbors, did not signify a continuing relationship between the families.

The farming enterprise of Daeng and Nawm was reliably profitable. It not only provided them with year by year security and one of the higher standards of living in the region; it also made possible a great expansion of their capital base (that is, their landholding), and enabled them to provide urban education beyond the minimally required four years (now seven) for most of their children. In so doing, they had in effect made a critical—and definitely non-traditional—decision as to how at least some of their wealth would be passed to their children: instead of giving them land at the time of their marriage or a substantial bride price which would in effect purchase rights in the land of another family (either of which Daeng and Nawm could have well afforded for all of their children), they chose to invest in more and better quality education than could be obtained locally, with the purpose of launching some of their children upon urban based civil service careers. 2) Rym was a young man in his late 20s when we knew him, who lived in the household of Daeng and Nawm. Although they were thought to be relatives (*jaad*), the link was so distant and obscure that no one could specify it (this degree of relatedness, which never signifies in and of itself socially active kinship ties, obtains between many or most residents of the region). Nevertheless, Rym was socially very much a member of the family: a child (*luk*) to Daeng and Nawm, and an older sibling to their children. Daeng and Nawm had housed him since he was a child, and had seen to all of his material needs with the same generosity they extended toward their own children. Rym in turn, had since adolescence labored willingly and ably on their lands, and had been a loyal and productive member of the household. He was uniformly well liked by his "younger siblings," who extended an easy but genuine respect to him and allowed themselves to be guided by him. He had substantial independence and responsibility within the household, albeit within the orbit of dominance of Daeng and Nawm. The ashes of Rym's parents reposed in urns next to the house shrine, and Daeng and Nawm provided a substantial bride-price for Rym at the time of his marriage, further signifying that they accepted him as a member of the household in good standing.

Although Rym was the only member of his family living with Daeng and Nawm while we resided in Banoi, in fact the relationship between the two families had existed for two generations. Rym's parents, themselves landless, had lived for many years with Daeng's parents' family, in much the same manner as Rym subsequently lived with Daeng and Nawm, and Rym and his siblings were born into a situation in which their family existed in the shelter of the informal patronage provided by Daeng's parents' family. But if the relationship was in one sense that of patron and client, it was also heavily stamped in the kinship idiom. For although perpetuation of the relationship depended upon exchange—work and services from Rym's family in return for economic security and refuge from hardship—so also did it depend upon the maintenance of the easy informality, which is not at all incompatible with deference, that characterizes harmonious households.

Although we learned of only two active relationships of this sort in and around Banoi during the 1960s, this type of situation is reminiscent of arrangements that were almost certainly more popular earlier in the century; and it is reminiscent as well of the types of arrangements that at least in many cases must have existed under the auspices of bondage. Traditionally, even before the changes described in these pages got underway, each generation produced its quota of landless villagers, albeit at nowhere near the levels of the 20th century.[29] The common practice of fragmenting inheritances produced this circumstance, and anything—misfortune, profligacy, or whatever—that dissipated resources and thereby conduced to subsequent disinheritance of descendants abetted it. Some landless families doubtless elected the pioneer option as the rejoinder to their dilemma. Others, however, sought less challenging alternatives. The latter, conservative choice made by a villager in the 19th century would often have led him along with his family if he had one into bondage. Following the demise of bondage at the close of the 19th century, a conservative disposition would have led the landless villager to seek a patron family with whom he and his dependents might find refuge in return for services performed and loyalty extended. In neither case would it have been a foregone conclusion that a satisfactory relationship could be established; in both cases the would-be client would have looked not just to the wealth but also and especially to the character of his potential patrons. What he learned on this score—for not everyone could expect to be as fortunate in their benefactors as were Rym and his family—might well have been the decisive factor in determining whether he with his family would opt for clienthood or strike out on his own. If a relationship did ripen, however, whether between slave and master or client and patron, it is likely that it would have had many of the diffuse, kinship-like properties characteristic of the relationship between Rym and his benefactors. And if this is so, it is understandable that not just landless villagers but also many who subsisted only marginally and precariously on small holdings would have preferred clienthood, sometimes even bondage, to the dubious benefits of an insecure and perhaps impoverished freedom. 3) We knew Nanuu as a landless villager in her forties. She had her own house, in chronic disrepair, in which she lived with her husband and two teen-age children from an earlier marriage. The house and surrounding compound, including a pond with a much prized fresh water spring in it, was all that remained of an estate which in her childhood had been substantial. Her mother had died when she was a girl, however and her father remarried, with unfortunate consequences for Nanuu. The step-mother had adult children of her own from an earlier marriage, and her loyalty and affection was for them and not at all for Nanuu and her siblings. She managed somehow—by diabolical scheming and the use of a much despised form of love magic (*tam saneej*), according to Nanuu—to get control of the land from Nanuu's father, and all of it went to her own children and none to her step-children. Traditionally, before acute land pressure developed, this was one of the ways in which landlessness could occur: at the parent generation, a rival claim is established on the family estate, and at least some of the children are disinherited as a result.

29 Piker, "The post-peasant village..."

Nanuu had clearly been a loser in this kind of situation, regardless of the veracity of her love magic accusations. In the light of the total unavailability of unused land in Nanuu's adulthood, and the very high scale price of paddy land due to its commercial value, there was no possibility that Nanuu along with other landless villagers of her generation could somehow acquire new land. She, and they, were irretrievably consigned to a landless existence.

For some twenty five years—the duration of her adulthood—Nanuu has been coming to terms with this exigency and making ends meet, however tenuously. It will be instructive to see how she manages. Her subsistence has depended upon two types of employment, albeit their relative importance has changed over the years. First, she has taken whatever wage work she could find locally. As with other landless villagers, agricultural wage labor is by far the most significant for her. Since, however, her children are not old enough to work and she was for some years without a husband, Nanuu has felt obliged to take any and all other temporary jobs. Some of these, such as digging fish ponds for wealthier neighbors, involved despised kinds of labor and comparatively low pay, but her marginal existence left her no choice.

Second, Nanuu has attempted to cultivate informal patron-client relationships with a number of landed families, and over the years her efforts have produced some success. Would-be clients such as Nanuu do not, of course, make formal application to potential benefactors. Rather, one's candidacy is advanced informally and circumspectly, yet in most cases unmistakably. The would-be client, typically, finds some reason to be in the compound of a potential patron when work is being done, and simply pitches in and helps with no expectation of payment; or perhaps, if hired to do a specific job—such as cutting firewood—finds an opportunity to perform a service over and above the labor contracted for. If the candidacy prospers, an informal relationship will flower. Part of it will consist simply of new found conviviality between client and patron: the client spends leisure time in or around the compound of the patron, occasionally eats or drinks with the patron family, and casually lends a hand if some job needs to be completed. Both parties appreciate the congeniality of such encounters, and not insubstantial services for the patron family may materialize in this manner. Above and beyond this, however, the client receives tangible benefits, which may best be described in terms of probabilities: the probability that if the patron family has wage work to offer, the client will have first shot at it; a higher than usual probability that, should real hardship strike, the client may obtain a loan at low or even no interest from the patron (loans obtained locally command interest rates of up to 15% a month; and interest rates are likely to be higher for poor villagers who have no collateral than for well to do villagers); the possibility, in no circumstances great, that in times of hardship an outright gift of cash or rice will materialize; the loan of buffaloes and plows to cultivate small patches of public land to which landless villagers have access; and so forth.

Landless villagers often attempt to maintain themselves in more than one such relationship at the same time, since what they receive from any one will not in all likelihood be great. Nor are such relationships normally durable. Of the two patron-client relationships in this idiom in which Nanuu was involved in the 1960s, neither had been in existence for more than a year. By the same token, she told us of earlier patron-client relationships which had lapsed. To the extent, however, that landless villagers are able to establish such relationships, they thereby achieve an added and often significant increment of security. They also are aided in maintaining a self definition which, although no longer fully justified by the objective circumstances of their existence, is approximately continuous with the traditional image villagers held of themselves as rice cultivators and social equals; for the landless villager *qua* client subsists largely by employing the skills he acquired as a youngster working in his parents' fields, and the typical informality of patron-client relationships nurtures the conviction that the palpable differences between patron and client are of wealth but not of station in life.

The possibility for numerous, if tenuous and temporary, informal patron-client relationships of the type Nanuu has cultivated doubtless has contributed importantly to the ability of a large number of landless villagers to continue to reside in the village. It has also contributed to the persistence of the "peasant community" flavor of daily life in the village, although the manner of organization of agriculture taken to be distinctive of peasant societies and peasant communities is now for Banoi largely a thing of the past.

4) We knew Sin as a young, married, and landless villager who resided with his wife and young children in a house on a piece of public land on one edge of Banoi. The village headman (*phujaibaan*) had given him permission to build and live there, so it was legal for him to remain there indefinitely. He could, however, never own the land upon which the house rested. Unlike most landless villagers, Sin's parents had themselves been landless. So although Sin like other landless villagers, subsisted mainly by agricultural wage labor, he was unusual in that he had learned the requisite skills as a wage laborer, and not while working in the fields of his parents.

Whereas Nanuu had never attempted to live anyplace other than Banoi, Sin had given Bangkok a serious try. With a friend he had lived there for two years working at whatever jobs he could get. Invariably, these proved to be low paying "construction" jobs, usually digging sewers or working as an unskilled laborer on the new roads being built in and around Bangkok. At least during the early 1960s, many villagers—both men and women—who tried the Bangkok labor market found jobs only of this sort. Mostly, Sin's jobs lasted no more than a few weeks, and he often waited almost as long for the next job. The work invariably was hard, dirty, and miserable; and the pay, although more than he could make on a per diem basis in the countryside, was barely sufficient to allow him to subsist in filthy and crowded workers' quarters. After two years of this, with no chance to save and no cash surplus for recreation, Sin had had enough. Realistically perceiving that he had virtually no chance of ever getting a better job in Bangkok, he decided to return to the countryside where, even if he were poor which he almost certainly would be, he at least would not be ground down by overcrowding, filth, and daily backbreaking labor. Upon his return to Banoi, he lived briefly with his parents. When he married the daughter of another landless family shortly thereafter, he erected a small house in his parents' compound and, when he received permission to live on public land, dismantled the house and reassembled it there.

On first inspection, it appears that Sin and Nanuu make a living in much the same manner: both depend primarily upon agricultural wage labor, and both supplement income from this source with occasional odd job employment. There is, however, one important difference: whereas Nanuu has had some success in cultivating informal patron-client relationships, Sin has had none. It is not that he fails to perceive their desirability; on a number of occasions we overheard him express envy for those who are more successful than he has been in this regard. Nor does he fail to try. For some reason, however, his candidacies for client status have never met with success. It may be in part due to the fact that he, like many young men, has a reputation for "hot heartedness," a proneness to lose his temper and get into fights or otherwise be unreliable. It may also be that his assumption of the traditional village identity is less convincing because, for life history reasons, he has not directly experienced the full range of circumstances in which the identity is rooted. Whatever the reasons, however, Sin is representative of a small but rapidly growing group of villagers who have not known, either in their own lives or through observing as children the lives of their parents, what it means to subsist as a cultivator of family lands. Correlatively, Sin is a member of a growing group of villagers whose subsistence depends entirely on contractually arranged wage labor, and whose access to work is not significantly eased by social relationships which, if they are not (and usually they aren't) between kinsmen, are nonetheless defined by an idiom that is highly reminiscent of kinship.

IV. Conclusions: new forms of social stratification in the countryside

Thailand, like all state organized societies, has been highly stratified for its entire national history. The *sakdina* system, outlined above, both codified and solidified the traditional patterns of social inequality between classes of people. More recently, beginning in the 19th century, the pattern of stratification in Thailand has taken on yet added dimensions and complexity as the forces of "modernization" described in these pages formed and took effect. The ethnographic present represented in this paper—the 1960s—reflected an advanced stage in the development of this process: the peasant society situation which had persisted in its essentials to the end of the 19th century in the Central Plain had given way to the onslaught of commercialization and its corollaries, and erstwhile peasants now stood in new and changing relationships to the broader society and, significantly, to each other.

The main purpose of this paper has been to trace the historical development of this last named circumstance, namely, the emergence of of new patterns of relationship among villagers themselves. The focus has been the village, and those who remain there; the manifold changes occurring in other sectors of Thai society, many of which involve one-time peasants, have not concerned us. And the major conclusion of this paper is that definite patterns of social stratification, predicated upon differential access to the wherewithal for subsistence and the occupational implications thereof, are emerging among villagers, whereas previously in Thai history villagers have stood as a single stratum in relationship to the other strata of the society.

That this development is taking place is obscured somewhat, both to the villager and the observer, because the daily run of village life continues to have a "peasant community" flavor. This in turn is due, first, to the fact that most landless villagers were born into landed families and as a result grew up with the traditional self definition; and, second, to the fact that the idiom of relationship between landed employer and landless wage laborer, although usually not in fact kin based, nevertheless is stamped in the kinship idiom, thus blurring the "objective" social class distinctions between them. The prognosis of this paper, however, is that this blurring effect will prove to be very much a temporary thing, and that "objective" economic conditions will soon succeed in transforming not only the social organization of agricultural production, but also and especially the self images of villagers.

Contributions to Asian Studies, Vol. IX

Opening a Frontier

The Expansion of Rice Cultivation in Central Thailand in the 1890's

DAVID B. JOHNSTON

Phillips Exeter Academy, New Hampshire, U.S.A.

THROUGH THE first half of the nineteenth century, agriculture in central Thailand was largely the business of self-sufficient peasant households. According to one account ("Needed", *Bangkok Times*, 28 Sept. 1909),

> ...each family had a small patch of paddy land, growing enough rice for home consumption. There was a garden growing cotton, which our womenfolk converted into good, lasting panungs and pahom [*phanung* and *phahom*: traditional items of clothing]. We grew tobacco for our own smoke, and we had fruits and vegetables. We made our own boats and carts, and we were happy.

It is, of course, difficult to support this image with documentary evidence. However, elderly villagers in central Thailand today still remember times when nearly every household produced its own clothing. Some villages grew their own cotton and spun their own thread. Others purchased cotton thread from market towns, but all seem to have performed the final stage, weaving the finished product, at home. (Interviews: H.S.K.; Q.S.Kh.; H.S.Q.). In addition, most farmers not only grew their own rice, but they milled it domestically as well. (Vincent 1874: 180; Sa-nguan 1959: 314–15; Young 1907: 209; Gerini 1912: 163; Child 1892: 145). And, to balance their diet, they caught fish for protein and planted various vegetables and beans to go along with the wild greens which were theirs for the gathering.

By 1900, this traditional self-sufficiency was rapidly disappearing, as farmers throughout central Thailand were being drawn into a commercialized economy geared to the production of a surplus of rice for export. The foreign demand for Thai rice was first clearly felt in the 1850's, and it increased rapidly during the following decades. The overall statistical dimensions of the Thai response to this demand have been documented elsewhere, most thoroughly by James C. Ingram. Between the 1850's and 1930's, rice exports increased 25-fold. Moreover, while only five per cent of the 1850 crop was exported, by 1900 exports represented about half of the total crop (Ingram 1971: 37, 8, 24). During these same years, imports of foreign products—most notably textiles —also increased rapidly (Ingram 1971: 112–35), as farmers began to purchase at market goods which they had previously produced domestically. Clearly, by the turn of the century, Thai farmers were heavily involved in the commercialized economy. They had sacrified a significant degree of self-sufficiency in order to focus their energies on production for the export market.

Ingram has also shown that the area planted in rice expanded rapidly during these years (Ingram 1971: 43–45), and it is to this phenomenon that the bulk of this paper will be devoted. Increases in exports were, to a large extent, dependent on the expansion of the cultivated area, and, in order to facilitate this expansion, it was necessary to provide access to previously isolated and unsettled regions of central Thailand. By looking briefly at the manner in which these areas were opened up, settled, and brought under cultivation, we may achieve a useful perspective on the Thai response to the changing economic conditions.

Our attention will focus largely on areas to the east of Bangkok where the pattern of development has differed significantly from that of most of central Thailand. Whereas most areas experienced gradual population growth and an even-paced transition from self-sufficient to commercialized agricultural production, this east bank region underwent a rapid and radical transformation from uninhabited grasslands to highly commercialized farm holdings within a period of just two decades. The region is, then, atypical, but it is nonetheless important to the history of the expansion of the central Thai rice economy. In the first place, it came to represent a significant share—estimates range as high as 25 per cent[1]—of the total rice land in central Thailand. More importantly, however, the east bank was the focus of agricultural attention in the 1890's. Those interested in agriculture during these years tended to follow closely, and become involved in, developments taking place on the east bank. As we study these developments, it should become clear that Thais at all levels of society, from princes to peasants, grasped eagerly the new economic opportunities of the time.

In the 1850's and 1860's, the increasing foreign demand for rice was supplied largely by paddy produced in the heart of the plain, centering on Ayudhya. Watered and drained by the Čhao Phraya and Pasak rivers, this area had been the granary of the kingdom, providing the surplus which sustained the court and its military and civilian bureaucracies, both in the Ayudhyan (1350–1767) and, now, in the Bangkok (1782–) period of Thai history. Despite this important role, the area was only partly cleared and settled, and it still contained much room for the expansion of cultivation during the nineteenth

1 Statistics on cultivated area were always reported by province or other administrative unit. Since these units frequently contained both east bank and non-east bank areas, and since the east bank was never considered as an independent administrative unit, it is difficult to estimate the cultivated land in the area.

In 1905, the provinces of Samut Prakan, Nakhon Khuankhan, Thanyaburi, Minburi, and Chachoengsao, all of which lay almost entirely within the east bank, contained 1,264,996 *rai* of rice land. In addition, portions of other provinces, including Ayudhya, Bangkok, Nakhon Nayok, and Chonburi were on the east bank, and it is likely that their inclusion would push the total figure to something near 2,000,000 *rai*. At the same time, Ingram (1971: 44) has shown that total cultivation in central Thailand was on the order of 6,800,000 *rai*.

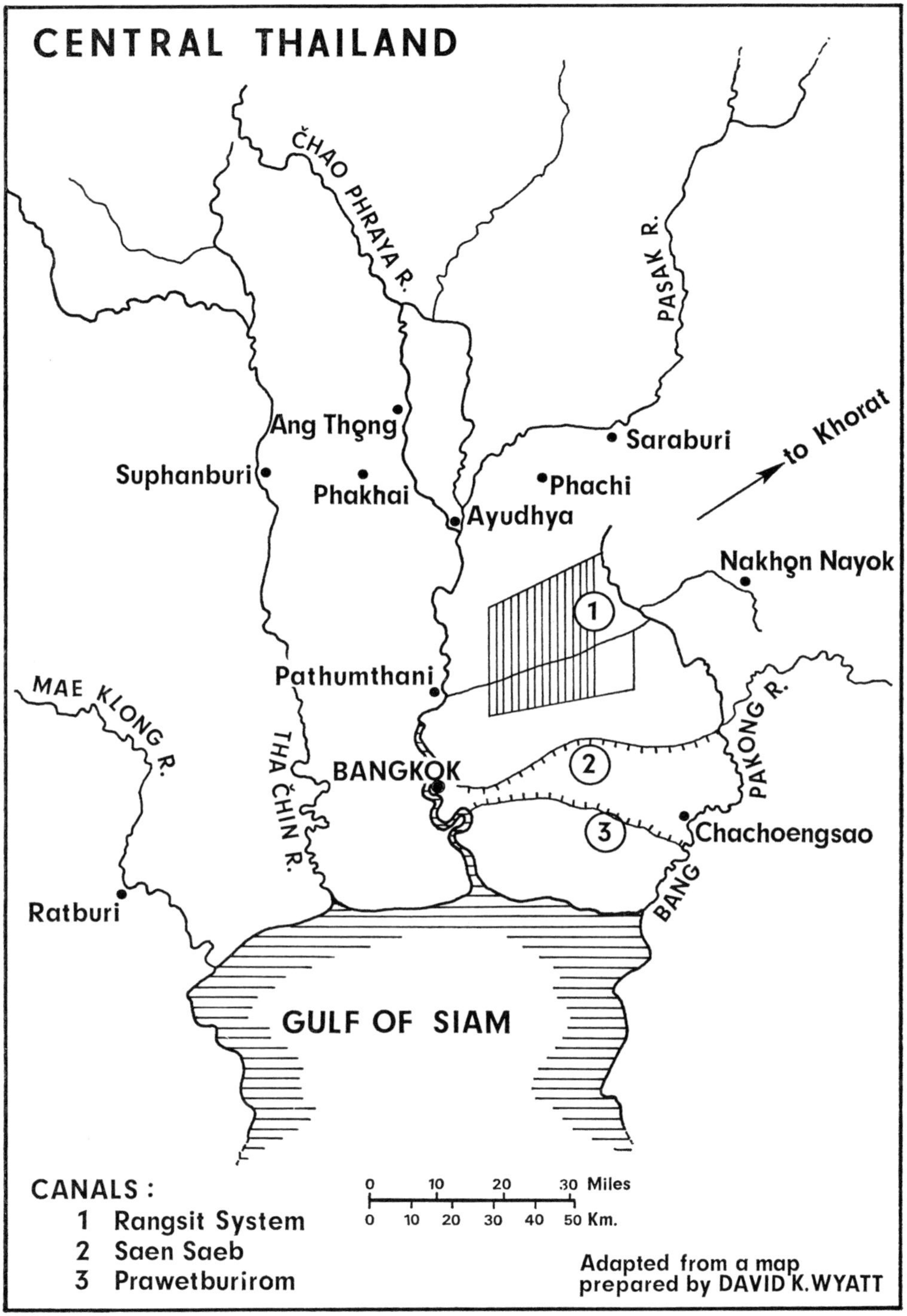
CENTRAL THAILAND
ČHAO PHRAYA R.
PASAK R.
Ang Thǫng
Suphanburi
Phakhai
Phachi
Saraburi
Ayudhya
to Khorat
Nakhǫn Nayok
Pathumthani
MAE KLONG R.
THA ČHIN R.
BANGKOK
PAKONG R.
Chachoengsao
BANG
Ratburi
GULF OF SIAM
CANALS :
1 Rangsit System
2 Saen Saeb
3 Prawetburirom
0 10 20 30 Miles
0 10 20 30 40 50 Km.
Adapted from a map prepared by DAVID K.WYATT

century. As the impact of the foreign demand was felt, and as farmers reacted to the new opportunities of commercialized rice production, the amount of surplus land began to dwindle.

Unfortunately, we have no series of data on population, cultivated land, or rice production which might enable us to follow this process closely. We must rely on indirect evidence, and on the impressions of contemporary travelers and the recollections of elderly Thais and their descendants. For instance, we know that, for several decades after the removal of the capital from Ayudhya to Bangkok in 1767, the Ayudhya provincial administration continued to govern through the same four districts (*khwaeng*) which had existed when Ayudhya had been the capital. Since a *khwaeng* was designed to govern only a limited population, and the practice was to divide one *khwaeng* into two if that limit were exceeded, we may assume that so long as Ayudhya contained four *khwaeng*, its population had not substantially increased. In fact, after the sacking of the capital by the Burmese in 1767, it almost certainly decreased. Then, however, sometime during the third reign of the Bangkok period (1824–1851), two new *khwaeng* were established in Ayudhya (Boranburanurak 1907: 117–19). This suggests that the population may have begun to increase somewhat earlier than the development of a strong foreign demand for Thai rice. The creation of additional *khwaeng* (later known as *amphoe*) in Ayudhya in 1895, 1903, and in the 1910's reflected continuing population increase and expansion of cultivation (Kachorn 1970: 12–13; Boranburanurak 1907: 119).

Elderly Thai living in Ayudhya today offer recollections which support this picture of expansion. By the 1880's, it seems, there remained in Ayudhya little unclaimed land suitable for expansion. In Phakhai, a market town and district administrative center northwest of Ayudhya, it is recalled that by this time new settlers, and refugees cast off from households grown too large for their holdings, were forced to migrate westward, in the direction of Suphanburi, to find unclaimed lands to clear and plant (Interviews: Ph.N.Ph.; H.N.Ph.).

To the east and west of this Ayudhyan heartland, there existed less well settled areas which remained in the nineteenth century, as they had been in the Ayudhyan era, quite peripheral to the core area. It is somewhat puzzling that, as the older areas of settlement offered decreasing opportunities for expansion, farmers did not expand their cultivation into these areas. To some extent, of course, they did. That this extent was limited, however, was due to two factors. First, the areas themselves were not ideal. To the east, south of the Pasak River and stretching eastward to the hills leading to the Khorat plateau, the slightly higher-lying lands received less regular flooding from the rivers, and drought was a constant worry. Moreover, the proximity to the Dong Phrayayen mountains, with their widely feared fevers, and the presence of numerous, harassing herds of elephants tended to retard the pace of settlement in these districts (Interviews: S.Ch.L.; H.S.K.; K.S.K.; Q.S.kh.). To the west, obstacles to settlement seem to have been less substantial, consisting primarily of a vulnerability to excessive flooding.

The settlement of these peripheral areas was also slowed by a second factor:

the sudden attraction presented by areas to the south as they were opened to cultivation during the final decades of the nineteenth century. This southern region can hardly have been considered even peripherally important to the economy of the Ayudhyan state. Containing strategically important outposts, it was largely an unihabited wasteland of jungle grass. But in the late nineteenth century, favored by proximity to the new capital at Bangkok and offering, in its virginity, a seemingly limitless potential, this southern frontier began to attract and absorb the energies of those Thais who sought to participate in and profit from the expanding economy.

The major obstacle to settlement and cultivation of the southern frontier was its inaccessibility. Farmers accustomed to the proximity of waterways—both natural rivers and manmade canals—and to the ease of transportation and communication which they offered were reluctant to embark on the settlement of this frontier which, in the long stretches between the major rivers, was transversed by few sizable streams. Before these areas could be well settled, therefore, it was necessary that canals be cut through them, connecting the rivers which flowed from north to south. The canals would enable farmers to enter the frontier and settle its lands, and they would also ensure that they would be able to get their surplus of paddy to market. Moreover, because rainfall was frequently insufficient for the needs of the rice crop, farmers could hope that the canals would facilitate the inundation of their lands: they would spread the rivers' flooding further into areas which might otherwise remain dry (NA Ag: N.K.: 86, 3839, Sai Snidvongs, Suaphan Snidvongs, and Patibatratprasong to Surasakmontri, 24 May 1894: 1, 7). The digging of the canals which opened this southern frontier was a crucial step in the expansion of the rice economy, and one at which we may now look in greater detail. For a number of reasons—relating both to historical conditions and to the thoroughness of documentation—it is easier to focus on the areas to the east, rather than the west, of Bangkok.

The traditional Thai state had been interested in canal excavation primarily for strategic and defensive reasons. Saen Saeb Canal, for instance, had been dug in the 1830's in order to facilitate communication between Bangkok and its troops fighting in Cambodia (Wongsanupraphat 1941: 133). Such canals often cut through unfarmed grasslands and they therefore did open new lands to cultivation, but this was never their primary purpose. It was not until the 1860's, when Thais began to recognize the potential economic value of their undeveloped plains, that canals were dug with strictly economic goals in mind. During the 1870's and 1880's, a number of land development projects were undertaken by both private entrepreneurs and public officials. Some, such as Prawetburirom Canal, were important beginnings, but it was only in the 1890's that these projects began to play a major role in the expansion of the rice economy.

The most important project was that of the Siam Land, Canals, and Irrigation Company. By the late 1880's, King Chulalongkorn was among those who had concluded that further agricultural expansion required additional canal development. Among those privy to such royal opinion was Phraongčhao Sai Snidvongs. Prominent member of an important branch of the royal family, Sai was the King's personal physician, an influential officer in the Navy, and proprietor of a substantial shipping business. Sai and the King apparently agreed not only about the need for further land development but also on two other points: first, that the royal treasury lacked the funds to finance the needed projects; and, second, that private entrepreneurs would not undertake them without clear assurances that they would be able to recoup the heavy investments which would be necessary. As the story goes, Sai developed a concession scheme; formed the Siam Land, Canals, and Irrigation Company (*Bǫrisat khut khlǫng lae khu na sayam*); and requested and recived royal approval for the concession (NA 5/1 N.3.2 K/2, Rụthičhakkamčhǫn, "Čhotmaihet"). Joining him as partners in the company were Joachim Grassi, an Italian architect and proprietor of a construction firm in Bangkok, and two others, Phra Nanaphitphasi (Chün) and Luang Sathǫnratchayut (Yom), of whom little is known (NA Ag: K.Kh.: 34, 791, "Čhaengkhwam", 25 March 1891; and "Irrigation in Siam," *Bangkok Times*, 15 Jan., 1903, p. 23).

The contract signed in January, 1889, gave the company a virtual monopoly on canal and land development projects throughout the kingdom for the next twenty-five years. Under the original terms, both the government and other private developers would be required to inform the company of any proposed land developments. Only if the company waived its privilege could the other party proceed. Further, the company could waive its right to manage and develop a project but still insist that it be contracted for the actual digging of canals; if its bid were lower than or equal to that of other contractors, it would receive the contract. The company was to be granted all land in the areas it opened up, to distribute or utilize as it saw fit. On the assumption that the company would choose to sell most of the land in its projects, the contract reserved to the government the right to regulate the prices at which sales could be made. Finally, the company agreed to pay a yearly fee, equal to twenty per cent of its profits, to the government (NA Ag: N.K.: 91, 3967).

The grand scale of the company's plans was clear from the outset. Deciding almost immediately that canal digging by Chinese wage labor was slow and, in the long run, more expensive, the company sent away to Europe for the first modern dredging and excavating machinery to be used in Thailand (NA Ag: N.K.: 86, 3833, Grassi to Surasakmontri, 5 May 1892: 2). Mapping and surveying were begun, and plans were made to lay out a criss-cross pattern of canals which would cover the whole of low central Thailand on both sides of the Čhao Phraya River. The first of these, Rangsit Canal, was the main artery of what came to be known as the Rangsit system. Running between the Čhao Phraya and Bang Pakong rivers to the north of Saen Saeb Canal, it was approved by the government in 1890, but the next two years seem to have been devoted less to canal dredging than to contract squabbling. Disagreements between the government and the company were to plague the latter throughout its checkered history. In 1890–91, it was upset by what it considered the obstructiveness of the Minister of Agriculture, Čhaophraya Phatsakǫrawong (Phǫn Bunnag) who, the company charged, was interested less in land development than in the land speculations of his relatives (NA Ag: N.K.: 86, 3833, Grassi to Surasakmontri, 5 May 1892: 6–7; Smyth 1898: I, 53–4). Not until Phatsakǫrawong had been moved to another position and Grassi's interest in the company bought out by another European did the work begin to progress satisfactorily.

By September of 1892, Čhaophraya Surasakmontri (Čhoem Saengchuto) had replaced Phatsakǫrawong as Minister of Agriculture, and was prepared to approve the company's request for permission to slice some forty canals to the south and north from Rangsit Canal. The plans were briefly snagged, however, by the problem of Grassi's legal status. Registered at the French consulate, he refused to place himself under Thai law. The government was wary of his right to claim extraterritorial privileges and threatened to abrogate the contract (NA Ag: K.Kh.: 34, 791, King to Surasakmontri, 11 Oct. 1892). Grassi refused to comply with government requests, but he did allow his interest in the company to be purchased by Erwin Müller, an Austrian partner in the B. Grimm Company in Bangkok. A long-time resident, Müller bore the Thai title of Phra Patibatratprasong, and brought to the company his apparently ample supplies of both capital and managerial ability. Within the year, Müller came to share a dominance in company affairs with Mǫmratchawong Suaphan Snidvongs, eldest son of Phraongčhao Sai. Known to the foreign community as Dr. Yai, Suaphan had recently returned from medical school in Scotland. Intolerant of the slow workings of the bureaucracy and the formality of court life, he shunned an official career and chose rather to combine his varied business interests with a little gentlemanly farming and a rural medical practice.

By October, 1893, Suaphan and Müller had submitted an extended proposal for east bank development. To the existing plans, they added two more east-west canals between the Čhao Phraya and Bang Pakong rivers, one running along the northern and the other along the southern reaches of the system. The project would eventually include a total of 1720 kilometers of canals (NA Ag: N.K.: 86, 3836, Suaphan and Patibatratprasong to Surasakmontri, 7 Oct. 1893: 1). This expansion of the system seems not to have been more ambitious than circumstances warranted. Demand for accessible rice land was strong and remained so into the new century. During 1896, applicants requested twice as much land as the company was able to make available (NA Ag: N.K.: 91, 3967, Patibatratprasong to King, 22 Nov. 1896). In 1898, demand for company lands was "greater than ever" ("Local and General", *Bangkok Times*, 9 April 1898: 2). By 1899, the new lands being opened in the northern sections of the concession were able to satisfy the requests of but one-third of the applicants (NA Ag: B.Kh.Kh.: 1, 12, Suaphan to Thewet, 18 Nov. 1899), and in 1900 the company continued to find it impossible to meet this popular demand (NA Ag: B.Kh.Kh.: 4, 46, Sai and Patibatratprasong to King, 14 June 1900).

Though unable to keep pace with this demand, the company's work was nonetheless impressive. Floods in the early 1890's slowed its operations, but by 1895 Rangsit Canal had been dug approximately ten miles in from the Čhao Phraya and, "... in the wake of the huge excavators, fields were being cleared and houses were rising on their posts" (Smyth 1898: I, 54). In the following year, the canal reached the Bang Pakong River and locks were put in place at either end. By 1900, some 500,000 *rai* (1 *rai* = .4 acres) had been made ready for cultivation, and when most of the excavation was finished a few years later, estimates of the total area benefitted fell generally between 1,250,000 and 1,500,000 *rai*, although some observers claimed as much as 2,000,000 *rai* (NA Ag: B.Kh.Kh.: 4, 46, Sai and Patibatratprasong to King, 14 June 1900: 1; NA 6 K.5/10, "Rụang," appendix 8; Van der Heide 1903: 82; "Klong Rangsit District," *Bangkok Times*, 23 Aug. 1907: 3).

The Rangsit system was the largest and most successful of the land development projects of the 1890's, but it was not able to satisfy the constantly increasing demand for land. As a result, numerous other projects were proposed and undertaken, both by the Siam Land, Canals, and Irrigation Company, and by other entrepreneurs. Without going into the histories of these various projects, we may note briefly that they were initiated by a wide variety of people, including low-ranking nobles, provincial governors, and local land tax assessors; Chinese traders and Vietnamese soldiers; and two high ranking princes, royal half-brothers of King Chulalongkorn.[2] In addition, these projects were not limited to the east bank areas of the southern frontier, but were undertaken in areas to the west and north, wherever uncultivated and unclaimed lands could be found. Some of these projects succeeded, a number were conspicuous failures. The important points for our purposes are, first, that they opened a wide expanse of land to new cultivation and, second, that they reflected the eagerness of those Thai with some access to capital to invest it in new and innovative ways. Significantly, however, these projects depended largely on popular demands for new lands, and our picture of the expanding rice economy must include a brief discussion of the manner in which the Thai peasantry settled and farmed the newly opened areas. Once again, it is easiest to focus our attention on the areas to the east of the Čhao Phraya River.

Any attempt to reconstruct the pattern of settlement of the east bank is hindered by a scarcity of statistical data. Indeed, because the government really extended its administration into the area only at the turn of the century, for the previous years we must rely entirely on the observations and estimates of private individuals. It is clear that the movement into the east bank took place largely during the last two decades of the century, and, generally, expanded from southern to northern districts within the east bank. Thus, while by the mid-1880's, a traveler along Prawetburirom Canal noted that much

2 I discuss these various ventures in my dissertation, "Rural Society and the Rice Economy in Thailand, 1880-1930" (Yale, 1975).

of the land along its banks was already cleared for cultivation (Gerini 1885: 1), extensive settlement along Saen Saeb Canal to the north did not occur until a few years later (Vincent 1874: 175). In 1888, according to a government surveyor, James McCarthy, there was "no cultivation" to the north of Saen Saeb Canal; rather, the area "was full of wild elephants which were dangerous" (NA Ag: N.K.: 62, 2483, McCarthy to Surasakmontri, 6 April 1895). Seven years later, McCarthy was again in the area, and he reported a vast change (NA Ag: N.K.: 62, 2483, McCarthy to Surasakmontri, 6 April 1895):

> I was astonished at the change that had taken place. In every direction the land was cleared of the heavy jungle grass which afforded shelter to wild elephants, which caused damage to those living on Klaung Senn Sep [Saen Saeb Canal].... The wild elephants were driven back and man entered into possession.

The area to which McCarthy referred stretched from Saen Saeb Canal into the southern portions of the Rangsit concession, and during the 1890's migration to the east bank reached its peak as over 100,000 people came to carve out their holdings in the Rangsit area ("Rice Milling," *Bangkok Times*, 20 June 1902: 4). By 1902, the scale of migration into and settlement of these east bank areas necessitated the establishment of two new provinces (*muang*) to facilitate government administration there, and in 1907 the population in northern portions of the concession had expanded sufficiently to require the creation of a new district (*amphoe*) (NA 5/1. N.3.2.K/67, Phitakthepthani to Suranarotthaseni, 6 May 1907).

The general outlines of the migration and settlement are easier to portray than the processes and patterns through which it was accomplished. Some migrants were probably motivated by the "push" of circumstances rather than by the "pull" of opportunities. According to Lucien Hanks, (1972: 99), there was a "trickle of persons displaced from expanding Bangkok," and one elderly Thai suggested to me that, with the abolition of slavery in the 1890's, many debt bondsmen were cast out of secure circumstances and forced to seek an independent livelihood (Interviews: H.S.Q.). But, on the whole, migrants to the east bank were acting positively, seeking their fortunes in an expanding economy. Some came from Bangkok, and others from as far away as northeastern Thailand, either bringing their families with them or taking wives in their new homes (NA 5/1.N.3.2.Ch/9, "Rai-ngan," March 1910: 16–17; NA 5/1.N.44.1/40, Suranarotthaseni to Sukhum, 31 Dec. 1907; "Irrigation Problem," *Bangkok Times*, 16 Aug. 1899).

The majority, however, appear to have come from rural areas in central Thailand, and of these we may assume that Nai Čhun was probably typical. Throughout the 1880's, Čhun had experienced many difficulties with the rice land he farmed in a district to the west of Bangkok. The supply of water was less dependable than the ravages of various pests, and he rarely harvested more than enough to feed himself and his family. Then, a friend heard of the still unclaimed lands between Saen Saeb and Rangsit canals, and of the reportedly abundant crops they could yield. With the friend and several others, Nai Čhun decided to move to the area in 1890. In his first year there, Čhun cleared and farmed a plot of about five *rai*. By 1893, he had expanded his holding to 30 *rai* and his abundant harvest included a surplus for trade (NA Ag: N.K.: 86, 3844, Nai Čhun to Siam Land, Canals, and Irrigation Company, 18 May 1898).

Much, probably most, of the land along Saen Saeb Canal and between it and the southern border of the Rangsit system, was brought under cultivation by small scale farmers like Nai Čhun. Owning usually between 50 and 200 *rai*, they conform to the accepted view that the expansion of rice cultivation in Thailand was carried out by small scale peasant proprietors, and that the role played by large landlords and moneylenders was a strictly limited one. (Ingram 1971: 65–66). For most of central Thailand, this picture is accurate, but in unmodified form it fits the experience of much of the east bank quite poorly. In the first place, those who sought their fortunes in the east bank found themselves locked in battle with a frequently hostile natural environment. Life on this frontier was not easy. Clearing the land was arduous work, and a family had to struggle to support itself before the first harvest came in. If any grown member of a household were stricken with fever or other disease, the loss of his labor even temporarily might foredoom the whole effort to carve out a family homestead. Once the crop was planted, it was vulnerable to numerous dangers. It might be trampled by a herd of elephants (NA 5/1. N.3.2.K/12, Rụthičhakkamčhọn to Naret, 3 Oct. 1902: 5–6), for instance, or fall victim to excessive or insufficient rainfall, to plant diseases or animal pests.

In the old core areas of the kingdom, where settlement expanded more gradually and where newly claimed areas bordered on long-settled ones, those whose efforts to expand cultivation were visited with misfortune might easily retreat to the household or village from which they had tentatively broken away. In the new frontier of the east bank, such retreats, while certainly possible, were more difficult and less common. Moreover, unable to rely on support from established neighbors, farmers on the east bank more frequently were forced to borrow in order to finance their operations (*Bangkok Times*, edit., 30 June 1905). Some possessed marketable commodities which they could put up as security for their loans, but others could offer only the promise of themselves or family members as debt bondsmen. If the loan were not repaid, or interest payments not maintained, the "mortgaged" individual was taken into the personal service of the creditor as a slave (*bao*; *that*) (NA Ag: N.K.: 88, 3911, petitions to Surasakmontri, August 1896). Most debts were probably incurred to cover cultivation expenses in the first years of settlement, but those who settled within the boundaries of the Rangsit concession also had to make payments for their lands. As a resident later observed, many lacked the capital to purchase the lands, and

> ...so had to borrow at 15 per cent. Many of these had in time to sell out to people with capital, and either then stayed on and rented a piece of land, or they took up new land on other canals that were [*sic*] still vacant. ("Causes," *Bangkok Times*, 25 Sept. 1909).

A similar fate must frequently have befallen those who borrowed for other reasons as well.

The life of a small scale farmer was, therefore, less secure on this east bank frontier than in areas of older settlement. At the same time, the opportunities and potentials of the area, and of the entire rice economy at the time, attracted the attention of men of greater means and broader ambitions. Some apparently operated purely as land speculators. Although it is difficult to find much specific detail about this group, we know that several of them purchased land from the Siam Land, Canals, and Irrigation Company in 1897. They each claimed rather large parcels of land, and requested that the parcels be divided and registered (apparently for re-sale) as separate plots (NA Ag: N.K.: 91, 3977, Sai and Patibatratprasong to Alangkan, 28 May 1897). Some years later, a landlord argued in the columns of the *Bangkok Times* that these speculators reaped profits of three to four hundred per cent in the Rangsit district, buying from the company and selling a few years later at prices inflated by the "great land boom" ("Causes," *Bangkok Times*, 25 Sept. 1909). Another source added that the speculators were "Bangkok people... [who] bought and sold for sport" (NA 6 K 5/4, "Rụang": 70–71).

Less elusive than the speculators are the large landowners, one of whom freely expressed his pride in the role he and others had played in the development of the Rangsit and Saen Saeb areas in the following passage:

> ...this immense tract of land... was a wild waste from the beginning of time, which was of no benefit to anyone.... Then a company very cleverly obtained a concession to excavate canals, and by doing so got the land. *But it is we who have paid for those canals, and have sunk our capital in trying to open up this district,* and by our united efforts and capital have collected together such a large population.... *How long would it have taken the Rasdorn* [*ratsadon*: common citizenry] *having no capital, and no help from the Government, to develope* [*sic*] *this district?* ("Causes," *Bangkok Times*, 25 Sept. 1909; emphasis in the original).

Large landowners could manage their holdings in two basic ways. Some utilized the labor of their debt bondsmen, other dependents and, when necessary, hired workers to clear and cultivate their lands. When the governor of Ayudhya purchased 720 *rai* of land to the north of Rangsit Canal in 1894, for instance, he assigned a group of his bondsmen to farm it (NA Ag: N.K.: 88, 3889, Chaiwichit to Ministry of Agriculture, 12 July 1895). In the previous year, an untitled commoner purchased a modest 150 *rai* plot from the Siam Land, Canals, and Irrigation Company, and he too chose to develop his property with the labor of bondsmen (NA Ag: N.K.: 87, 3869, Nai Ked to Siam Land, Canals, and Irrigation Company, 29 June 1896). Another substantial landowner was Tat Snidvongs, nephew to Sai and cousin to Suaphan Snidvongs, who had purchased a choice piece of land on the banks of the main Rangsit Canal. His holding was somewhat unusual, he was proud to point out, because he managed its operation entirely by himself, living on the land and directing the labors of his hired hands (Tat 1963: 382–83).

Although the above examples suggest that large landowners could manage their holdings as single farm units, most chose to divide them into parcels and rent them to tenant farmers. By the 1890's, the reforms abolishing debt bondsmanship were beginning to take effect, and this meant that a decreasing number of landowners could command the manpower to farm a large plot directly. Moreover, most of the landlords were city people, and they were probably more comfortable leaving the farm operations to tenant farmers. They therefore sought to attract tenants to their lands from Phetburi, Ayudhya, Angthọng, "... and every other rice growing district where people had not sufficient land to cultivate" ("Government," *Bangkok Times*, 6 Aug. 1912).

The scale of a landlord's operations varied widely, as can be illustrated by reference to a few specific examples. On the one hand, a Chinese named Chün owned a small amount of land which he agreed to rent to a Thai farm couple in 1893. Of Chün's land, some 20 *rai* was already cleared, and he agreed to rent this for about 40 liters of rice per *rai*, which would normally represent well under ten per cent of the yield (NA Ag: N.K.: 88, 3911, Mọn to Surasakmontri, 13 Aug. 1896).[3] In addition, he allowed the couple to clear and cultivate, rent-free, as much of his adjacent grass land as they could. By 1896, the couple was farming some 200 *rai*, but the records do not say what the terms of their tenure were by this time.

At the opposite end of the landowning scale, we may point to the operations of the Lamsai Company. Although the activities of this company are very poorly documented, it is evident that sometime in the early 1890's it purchased about 9000 *rai* from the Siam Land, Canals, and Irrigation Company in the southeastern section of the Rangsit concession. The Lamsai Company proceeded to excavate small canals on its own within this holding, and to attract tenants with offers of rent-free lands during the initial years of cultivation. As the lands were developed, the company increased the rents gradually to a high of three baht per *rai*, or about one sixth of the gross income, in 1905 (NA Ag: Ph.P., R.N.: 22,446, Gears to Wongsanupraphat, 21 June 1910).[4]

Between the extremes represented by Chün's small holding and that of the Lamsai Company there fell the vast majority of east bank landlords who rented their lands out to tenant farmers. Without going into the economics of tenantry and other agricultural holdings, we may note only that large-scale landowning and tenantry were important aspects of the social landscape on the east bank frontier. Although we lack statistics on the subject, and therefore are unable to state what proportions of east bank lands were developed by small scale owner-operators, by bondsmen and other dependents, and by tenant farmers, it is clear that these proportions differed here from what they were in areas of older settlement. Whereas in the latter owner-operators were

---

3 The most frequent contemporary estimate placed the normal yield at .33 *kwien*, or something over 1000 liters, per *rai*.

4 In 1905, according to data reproduced by Dilock (1908: unpaged appendices), prices in the area averaged about 54 *baht* per *kwien*. A yield of .33 *kwien* per *rai* would have sold for 18 *baht*, from which the rent would have been subtracted.

dominant and tenant farmers quite rare, in the newly opened areas owner-operators were less numerous and tenants substantially more so.

The existence of a landless tenantry points our attention to another feature of rural society in the newly opened areas, the high rate of geographic mobility. The English diplomat who wrote in 1886 of rural Siamese that "it is seldom that they cross the boundary of their provinces," (FO 69/90, Archer, "Report": 2) was, strictly speaking, guilty less of inaccuracy than of anachronism. For, as a more acute Italian observer had already noted in the previous year, Thais living on the east bank were quite mobile. G. E. Gerini described the large expanse of land cleared for cultivation along Prawetburirom Canal, and went on to comment that the dwellings set up on these lands were not of the permanent Thai type, but were rather "simple bambu [*sic*] or straw huts, apparently erected to serve as a temporary asylum to the cultivators of the adjacent paddy fields." Gerini commented also on the absence of men in the families which inhabited these huts: "They emigrate soon after the harvest is gathered to seek work elsewhere, and return only in time to plough and sow their paddy fields" (Gerini 1885: 1). Ten years later, a land registration official from the Ministry of Agriculture found that this same impermanence of settlement frustrated his survey work. After the harvest, he and his assistants would conduct surveys in order to issue land ownership certificates and assess the land tax, but often neither landowner nor tenant could be found in the area (NA Ag: N.K.: 65,2740, Yothanatikan to Surasakmontri, 5 Dec. 1895). And again, in 1909, it was said that numerous tenants in the Rangsit area resided there only temporarily during the farm seasons; after the harvest, they returned to permanent homes elsewhere (NA 5 K. 3.1/13, Wisetphochana, "Rai-ngan," Oct. 1909: 22).

Others who settled more permanently on their holdings were nonetheless attracted to frequent, if not yearly, migrations within the east bank. At least through 1900, the majority of these consisted of farmers moving into the Rangsit concession. And of these, a large number were abandoning holdings along Prawetburirom Canal. In the 1880's, the Prawetburirom area had been a rich agricultural district, but a number of factors—silting of canals, damage inflicted by weed and animal pests, soil deterioration, and interference with the water supply caused by the excavation of the Rangsit canals, were specifically mentioned—contributed to a decline in the area's reputation at precisely the time when the Rangsit lands were being made available (Thailand 1900: 3). Those who moved from Prawetburirom Canal into Rangsit were joined by others from neighboring areas such as Pathumthani, and a poor crop in 1897 in certain districts along Saensaeb Canal induced over one hundred tenant households to abandon lands there in favor of those in the Rangsit district ("Wandering Cultivator," *Bangkok Times*, 1 May 1912: 18). Thus, population movements reflected the alternately waxing and waning fortunes of different areas on the east bank:

> Very different from the scenes of desolation in Klong Pra Wete [Prawetburirom Canal]

> is the appearance of the great Rangsit system. Here the value of land goes higher every year, new branch canals are constantly being made and the land opened up by them is at once taken up.... A constant stream of immigrants flows into the country watered by this canal, cultivators from the impoverished Klong Pra Wete and from the poorer lands on the west sice of the Menam [Čhao Phraya River], leaving their old fields for these new and better lands.... (Thailand 1900: 3)

This mobility persisted after the development of the Rangsit concession; in fact, an agricultural recession after 1905 resulted in sizable movement out of that area. According to one source, emigrants from the area sought new opportunities in Nakhọn Nayok, and Pathumthani, and some even returned to Prawetburirom Canal ("Wandering," *Bangkok Times*, 1 May 1912: 18).

Mobility of population and impermanence of settlement were, then, aspects of society on this frontier which distinguished it from older areas of cultivation in central Thailand. The areas differed also in the physical structure of their villages. Traditionally, Thai villages have tended to be of two basic types, known to anthropologists as "strip" and "cluster" villages (Anuman 1972: 306–07; deYoung 1963: 8–9). Strip villages consist of narrow lines of houses strung out along river or canal banks, or along roadways, with vegetable gardens and rice fields located behind the houses. Focused less on a village core than the common riverine frontage, strip villages in thickly populated areas tended to melt into a continuous belt of settlement. Cluster villages, located not on river banks but in the midst of rice fields, tended to form more natural units, set off distinctly from neighboring villages.

On the east bank, much of the settlement followed a new pattern in which farmers established their homes in the midst of their separate plots of land and not in villages of either the strip or cluster type. For administrative purposes, the government eventually grouped up to fifty such isolated households into what it called villages, and anthropologists have labelled them, appropriately, "scatter" villages (Anuman 1972: 306–07; deYoung 1963: 10–11).

Observers first commented on this new settlement pattern in the 1890's, and they tended to relate it to a general unrest, and a high level of crime on the east bank. In 1894, for instance, Suaphan and Müller of the Siam Land, Canals, and Irrigation Company noted that settlers in the concession "established their homesteads in isolated clusters wherever they wished," (NA Ag: N.K.: 86, 3837, letter to Surasakmontri, 14 May 1894) and in 1898 the *Bangkok Times* suggested that much rural violence in areas near Bangkok resulted from the fact that farm houses were "scattered about the fields at considerable distances from each other" ("Local and General," 4 April 1898). In the following year, a report submitted by W. A. Graham reached similar conclusions with regard to the Rangsit area:

> One necessary step towards protection of property might be undertaken at once. The traveller on Klong Rang Sit cannot fail to be struck by the entire want of system in the location of the houses. It appears that on arrival, a man plants his house on the bank of the canal on the spot which pleases him best, with absolutely no reference to Pu Yai Ban or Kamnan [village headmen and commune chiefs]. Here he gathers round him

his cattle and belongings and resides in splendid isolation with the result that, when some night his house is surrounded by a gang of robbers, he finds himself half a mile away from the nearest neighbour who could give him help. (Thailand 1900: 4)

Geographic mobility on the part of central Thai peasants was important to the development of the frontier areas, but of equal importance to east bank agriculture was the role played by seasonal agricultural laborers from northeastern Thailand. In the newly opened areas, farmers were able to acquire larger plots of land than they had previously farmed in the older areas of settlement. Since both their own households and the local communities lacked the manpower to farm the larger plots, farmers came to depend on migrant northeastern workers who, because of cultural and linguistic factors, were referred to in central Thailand as "Laos." Just as central Thai farmers needed the additional labor which the northeast could provide, so the relative poverty and the limited opportunities of the northeastern economy insured that labor on a central Thai farm was worth a Lao's while: the system, wrote Chulalongkorn, could be "of benefit to both parties" (NA 5 K. 3.1/11, King to Damrong, 17 May 1906).

Occasionally, a Lao worker would be hired to help out with one particular task, such as plowing or harvesting, but more commonly he would be contracted to work the whole of the nine or ten month agricultural season, from April or May to the following February or March. During slack periods, he was expected to work on non-agricultural tasks in the household. He usually lived with the cultivator and his family, ate meals with them, and was provided with clothes and other necessities. His wages varied according to local conditions and his age and experience. In Ayudhya, farmers today recall that wages could be as low as forty *baht* for the agricultural season, while in the Rangsit district they climbed to 120 *baht* in 1910. During the 1890's and 1900's, the average pay seems to have been about 80 *baht* for the season (Interviews: Ch.Q.N.; S.Ch.L.; K.S.K.; Dilock 1908: 102; Child 1892: 143–44).

A cultivator in need of additional labor might send an assistant to Khorat or some other northeastern province to engage workers, but this was practical only for large farms which needed twenty or more workers (Interviews: K.S.K.). More frequently, northeasterners in search of work would themselves come to central Thailand, arriving at the beginning of the planting season in April or May. Traveling in groups of up to one hundred and more, they walked together from village to village. One or two would leave the group at each stop to join labor-short households, until eventually all had found work. After 1900, workers would make the last leg of their journey by train and they often found eager cultivators waiting for them at railway stations in Saraburi, Phachi, Rangsit and other commercialized districts (Interviews: Ch.Q.N.; S.Ch.L.).

Each group of workers was supervised by an individual known as the "captain" (*nai roi*). Almost all *nai roi* were originally northeasterners, although over the years they would build up a general familiarity with central Thailand, and several of them came to consider central Thailand their home. Each year, the *nai roi* returned to the northeast in search of agricultural workers. There, he would engage anywhere from 15 to 120 workers, and lead them to Khorat, taking up to three weeks to make the trip on foot from different areas in the northeast. From Khorat, they either continued their journey on foot or went the remainder of the way by train (NA 5/1 N. 3.2 K/36, statements of Nai Sikhanthamat, Nai Sing, Nai Le, Nai Un, and Nai Phlọi). Once his workers were distributed to different households, the *nai roi* might have no contact with them until the end of the season. Then, with apparently little variation, the workers and *nai roi* would meet to begin the trip back to the northeast on the third day of the waxing moon in the third lunar month, which fell in either February or March. Throughout central Thailand, farmers hoped to finish their final operations before what they referred to as "the day the Laos leave" (NA 6 B. 2.1/38, Prachachip to King, 5 June 1911: 4).

I have found only two unsatisfactory estimates of the number of northeastern workers who came to work on central Thai farms. In 1901, an official in Thanyaburi province, which included most of the Rangsit district, reported that 1094 northeasterners were employed on farms in that province (NA 5/1 N. 44.1/40, Rụthičhakkamčhọn to Naret, 14 June 1902). In 1904, over 5000 northeasterners applied in Thanyaburi for the official papers which would allow them to return legally to their own provinces (NA 5/1.N.3.2 K/63, Tatsadakọnkoson to Naret, 6 Feb. 1907). It seems likely that this is only a fraction of those who actually came to work in Thanyaburi that year, and of course many more went to other provinces as well. Yet, without precise data, we can do no more than concur with the writer who estimated two decades later that "several thousand" northeasterners made this yearly migration (Phecharada 1931: 808).

By the first decade of the twentieth century, the grasslands to the east of Bangkok were no longer inaccessible and, to a large extent, they had been settled and brought under cultivation. Thais at all levels of society had played important roles in the development of the area, and, in so doing, they had demonstrated a willingness to abandon tradition in order to respond to new economic opportunities. The entrepreneurs who formed companies and vied for land development concessions, the speculators who gambled on increasing land prices, and the landowners who not only bought up but also retained and managed paddy lands had all, in their various ways, reacted to the challenges and opportunities of the time. Even more striking, however, was the role played by the actual tillers of the soil, whether small scale owner-operators, landless tenants, or migrant laborers. In each case, they showed a willingness to abandon the traditional self-sufficient economy in order to increase their farming profits; to tear themselves away from inherited family lands and set off in search of larger and more productive holdings; and, in settling in the "scatter" pattern, even to do without the security of the traditional village structure. During the 1890's, at least, it seems clear that a significant number of Thais concluded that it was in their interest to enter the modern, commercialized economy, and they were willing to sacrifice much of their traditional life in order to do so. And because they were responsible for the opening and settlement of the east bank frontier, they played a crucial role in the expansion of the commercialized rice economy.

REFERENCES

Interviews referred to in the text were conducted in central Thailand during 1972. The letters which appear in parentheses designate villages and individuals interviewed.

All references which begin with "NA" indicate documents in the National Archives in Bangkok. The accompanying letters and numerals refer to the organization of files in the archives.

References beginning with "FO" indicate documents in the Foreign Office Series, Public Record Office, United Kingdom.

References to Published Materials

1 ANUMAN, Rajadhon, *Phraya* (Yong Sathiankoset)
1972 *Kansụksa rụang prapheni thai lae chiwit chao thai samai kon* (An Investigation of Thai Traditions and The Daily Life of Thais in Former Times). Bangkok: Khlang Withaya.

2 BORANBURANURAK, *Phraya* (Phon Dechakup)
1907 *Tamnan krungkao* (History of the Old City). Bangkok: Aksonnit Press.

3 CHILD, Jacob T.
1892 *The Pearl of Asia.* Chicago: Donohue, Henneberry, and Co.

4 DEYOUNG, John E.
1963 *Village Life in Modern Thailand.* Berkely and Los Angeles: University of California Press.

5 DILOCK, Prinz von Siam
1908 *Die Landwirtschaft in Siam.* Leipzig.

6 GERINI, G. E.
1885 "A Trip to Khorat". *Siam Weekly Advertiser,* 2 May 1885: 1.
1912 *Siam and its Productions, Arts, and Manufactures.* Hertford, England: Stephen Austin and Sons, Ltd.

7 HANKS, Lucien M.
1972 *Rice and Man: Agricultural Ecology in Southeast Asia.* Chicago: Aldine Atherton.

8 INGRAM, James C.
1971 *Economic Change in Thailand 1850–1970.* Stanford: Stanford University Press.

9 Kachorn Sukhabanij.
1970 "*Krung si-ayudhya ratchathani tae boran* (Ayudhya, Ancient Royal Capital)". *Warasan Ayudhya,* III, no. 6 (June): 8–13.

10 PHECHARADA, Phraya.
1931 "*Rai-ngan pračham pi khọng monthon nakhọn ratchasima* (Annual Report of monthon Nakhon Ratchasima)". *Thetsaphiban,* XXXI, no. 11: 735–808.

11 Sa-nguan Ankhong.
1959 *Sing raek nai mụang thai* (First Things in Thailand). Bangkok: Phrae Phitthaya Press.

12 SMYTH, H. Warington.
1898 *Five Years in Siam from 1891 to 1896.* 2 vols. London: John Murray.

13 Tat Sihasaksanitwong Chumsai
1963 *Chokchata nai chiwit thi phọčhai* (Destiny in a Satisfying Life). Bangkok: Teachers' Institute Press.

14 THAILAND, Metropolitan Revenue Department.
1900 *Second Annual Report of the Bangkok Revenue Department, for the year 118.* Bangkok.
15 VINCENT, Frank, Jr.
1874 *The Land of the White Elephant: Sights and Scenes in South-Eastern Asia.* New York: Harper and Bros.
16 WONGSANUPRAPHAT, Čhaophraya (M.R.W. Sathan Sanitwong).
1941 *Prawat krasuang kasettrathikan* (History of the Ministry of Agriculture). Bangkok: Sophonphiphatthanakon Press.
17 YOUNG, Ernest
1907 *The Kingdom of the Yellow Robe.* 3rd edit. London: Archibald Constable and Co., Ltd.

Contributions to Asian Studies, Vol. IX

In Search of Land: Village Formation in the Central Chi River Valley, Northeastern Thailand[1]

CHARLES F. KEYES

University of Washington, Seattle, U.S.A.

1. Introduction

ONE OF THE MOST densely populated areas in rural Thailand today lies in the central Chi River Valley located on the Khorat Plateau of northeastern Thailand. The provinces of Mahasarakham and Rọi-et, which cover most of this area, have population densities which are exceeded in the kingdom only in the urban area around Bangkok. Despite the heavy concentration of population in the central Chi River valley, the settlement of the area has been quite recent, dating from the late 18th-century in the oldest settled part to the middle of the 19th-century in the most recently settled part. In this paper, I will focus specifically on one part of the central Chi River Valley, a part which includes the districts of Kantharawichai and Mụang Mahasarakham lying respectively north and south of the Chi. For these two districts, historical data exist which permit us to trace the processes of initial settlement and expansion of population in some detail.[2] These processes can be followed

---

1 A preliminary version of this paper was read at a symposium on "Population, Land and Structural Change in Sri Lanka and Thailand", held at the annual meeting of the American Anthropological Association, Mexico City, November 1974.

2 Primary historical information is available mainly from three chronicles: (1) "Phongsawadan huamuang monthon isan" ['Chronicle of the Provinces of the Northeastern Circle'] compiled by Mom Amorawong Wicit (1963); (2) *Phongsawadan phak isan* ['Chronicle of the Northeastern Region'] by Phraya Khatthiyawongsa (1929); and (3) *Prawat haeng phak isian lae mahasarakham bang ton* ['Some Aspects of the History of the Northeastern Region and Mahasarakham'] compiled by Bunchuai Atthakon (1962). Mom Amorawong Wicit, who was a Siamese official posted to the Northeast, compiled his chronicle in 1904 from sources obtained locally in the region. The recension of the chronicle used here includes some footnotes added to the original text which relate specifically to the history of Mahasarakham. Phraya Khattiyawongsa (Lao na Roi-et), the last of the traditional lords of Roi-et province, compiled his chronicle, presumably from family records available to him, in 1912. Bunchuai Atthakon, a descendent of the traditional ruling family of Mahasarakham and former Lord Mayer of Mahasarakham town, compiled his account in 1938 on the basis of family records as well as on such published material as appears in the the chronicle compiled by Mom Amorawong.

up to the present time through the analysis of yet other data collected during recent anthropological investigations.[3]

Initial settlement of Kantharawichai District by the forebearers of the present population began in 1836, and by 1845 the first settlements in the district were reported to have a total population of about 3,000 (Mọm Amọrawong Wicit 1963: 57). In 1970, the district had a total population of 50,544. First settlement of Mụang district Mahasarakham began sometime before 1850 and in 1865 the total population was recorded as being 9,000 (Mọm Amọrawong Wicit 1963: 73–74). In 1970, the total population of Mụang District was 107,046.[4] Assuming that the initial populations as recorded in the chronicles included all the peoples living in the areas now incorporated within these two districts, it is possible to obtain comparisons of the population densities as they existed at the time of original settlement and as they exist at present (as of 1970). Kantharawichai district, which today has a total land area of 254 sq. km., has experienced an increase in density of population from 12 persons per square km. in 1840 to 199 persons per sq. km. in 1970 while Mụang district Mahasarakham with a total land area of 770 sq. km. has experienced an increase from 12 persons per sq. km. in 1865 to 139 persons per sq. km. in 1970.

It is obvious from these figures alone that the increase in population which has occurred in the two districts of Kantharawichai and Mụang in Mahasarakham province has placed increasing pressures upon the land resources of the two districts. In this paper, I shall attempt to trace how the peoples in the area in question have coped with these pressures. In doing so, it is necessary to take into account not only their mode of ecological adaptation, but also the cultural conceptions of the relationship between land and labor which are held and have been held by those living in these two districts in the central Chi River Valley.

2. Initial Settlement of the Central Chi River Valley

In the districts of Kantharawichai and Mụang Mahasarakham there exist a large number of archaeological sites which attest to this part of the Chi River Valley having been the locus of a number of major centers of civilization which existed from about the 3rd to the 10th centuries A.D. In the 12th-century,

3 I myself carried out research in Muang district Mahasarakham in 1962–1964 and have returned for shorter trips during 1967–1968 and 1974. My research has been focussed primarily on the village of Ban Nọng Tụn, Tambon Khawao, Muang district, Mahasarakham. I am indebted to the Foreign Area Fellowship Program, the University of Washington, and the Ford Foundation for support of my research in Mahasarakham. Other anthropological research was carried out in 1967 by a number of researchers who made studies over a period of a month in villages in Muang and Kantharawichai districts. The work of these researchers (cited individually below) was carried out under the supervision of Dr. Toshio Yatsushiro, then of the United States Operations Mission to Thailand.

4 1970 population figures are taken from the report of 1970 census (Thailand. National Statistical Office, Office of the Prime Minister 1973:1).

as is evident from other archaeological remains, the area was incorporated into the Angkorean empire. With the decline of Angkor in the 13th and 14th centuries, the Khorat Plateau became something a political no-man's land, a character it retained until the 17th century. During this period between the 13th and 17th centuries, the Khorat Plateau appears to have become almost totally depopulated, and legends speak of the movement of people out of the region —presumably into Cambodia. The previous inhabitants of the Khorat Plateau did, however, leave a legacy for the new inhabitants of the region, for the sites chosen for new settlements were often ancient settlement sites, and the water tanks used by new settlers were not rarely ancient ones which were restored.

In the 17th-century, both the Siamese of central Thailand and the Lao of central Laos began to expand towards and onto the Khorat Plateau. Siamese settlers migrated into areas near the present town of Nakhǫn Ratchasima, but there they stopped and few, if any, Siamese moved further east or north until the 19th-century. Meanwhile, Lao from central Laos began a southward expansion which reached its furthermost extent with the founding of Campasak in southern Laos in the early 17th century.

Early in the 18th-century, peoples from Campasak began settling on the Khorat Plateau in the lower Chi valley. This expansion of the Lao was associated with the establishment of small "domains", known as *mụang*, which were ruled by "lords" (*cao*) and their kinsmen. In 1718, the first Lao *mụang* in the Chi valley—and indeed anywhere in the interior of the Khorat Plateau —was founded at Suwannaphum in present-day Rǫi-et province by an official in the service of the King of Campasak. This Lao noble, Can Kaeo by name, brought with him, so it is recorded, some 3,000 "subjects" (*phrai*) from Campasak (Phraya Khattiyawongsa 1929: 8; Mǫm Amǫrawong Wicit 1963: 26). About a half century later, a conflict between the sons of Can Kaeo led to the splitting of Mụang Suwannaphum and the creation of the new domain of Rǫi-et with its seat near the Chi River north of Suwannaphum. These two domains were incorporated into the kingdom of Siam following the reconstitution of the kingdom in the 1770's. In 1793, a third domain, Kalasin, was founded by a disaffected nobleman from the Lao kingdom of Vientiane who led a following including 4,000 men who were responsible for paying taxes and fulfilling corvée obligations (Mǫm Amǫrawong Wicit 1963: 41).

The three domains of Suwannaphum, Rǫi-et and Kalasin, while ruled by Lao lords (*cao mụang*), were all included within the kingdom of Siam as "outer provinces" (*hua mụang nǫk*). Their status differed from the Lao principalities of Vientiane and Campasak which were vassal states to Siam and differed from the status of the trans-Mekong domains which were under either Vientiane or Campasak. The relationship between the Lao and the Siamese polities altered sharply as a consequence of the Siamese-Lao war of 1827. This war began with an attempt by the rulers of Vientiane and Campasak to wrest their independence from Siam and, at the same time, to bring the whole of the Khorat Plateau under their sway. This attempt ended in utter failure as Siamese troops razed Vientiane to the ground, sent off members of the Vientiane royal family (including the King) to Bangkok as prisoners, and brought to an end, once and for all, the kingdom of Vientiane. The domains under Vientiane, lying for the most part to the east of the Mekong, were made directly dependent upon the Siamese throne. Campasak was reduced from a vasal state to a domain and was also stripped of most of its dependencies which were also made directly dependent upon Bangkok.[5]

The Siamese-Lao war of 1827 was a major watershed in the settlement of the Khorat Plateau, including the central Chi River Valley. Following the war, the Siamese court pursued the policies of encouraging (with force if necessary) Lao living east of the Mekong to migrate and resettle on the Khorat Plateau and of stimulating the proliferation of domains located upon the Khorat Plateau.[6] Both policies were designed to prevent serious challenges to Siamese authority being mounted by peoples living in the inaccessible areas lying to the northeast.

In 1836, the King of Siam ordered two of his commanders to raise a force among the populace of northeastern Thailand to go put down rebellions in what is today the Lao province of Khammouane (Cammon). The commander of the Siamese force pressured two of the local Lao lords, those of the domains of Khammuan and Khamkoet, to take their followers and to resettle on the Khorat Plateau in territory then belonging to the domain of Kalasin. One of these new settlements, that formed by the former lord of Khamkoet, was located at Tha Khọn Yang on the banks of the Chi River in what is today Kantarawichai District, Mahasarakham. This settlement is reported to have included a total of 2,859 people (Mọm Amọrawong Wicit 1963: 57).

In 1845, the King of Siam invested Phra Khamduan (or Lamduan or Khamkọn), formerly lord of Mụang Khamkoet, as the lord of a new domain called Tha Khọn Yang, Phra Khamduan was given various insignia of office and a new royal title and name as was done for every new lord of a domain on the Khorat Plateau before and after this event. The annual tribute levied on the domain consisted of 40 *hap* or about 60 kilograms of wild cardamons per year. The total number of "registered persons" (*tua lek*), that is those adult men responsible for paying taxes and subject to the corvée, was recorded as 407. Tha Khọn Yang was made a dependent domain to the larger domain of Kalasin, thus allowing the lord of Kalasin to claim a portion of the tribute for himself (Mọm Amọrawong Wicit 1963: 59).

Although nothing is said in the chronicles about where the population of Tha Khọn Yang actually settled other than in the village of Tha Khọn Yang itself, it can be presumed that most of the settlements were in the fertile areas lying to the north of the Chi in the vicinity of Tha Khọn Yang. Villages in this area remain distinctive today, their populations identifying themselves as

5 For a detailed discussion of the Siamese-Lao War of 1827 see Walter Vella (1957: 78–93).

6 On the policy of resettlement of peoples living east of the Mekong on the Khorat Plateau see Prince Damrong Rajanubhab (1951).

being ethnically Yọ rather than Lao. Yọ appears to be an ethnolinguistic variant of the dominant Lao culture followed by peoples in the Lao province of Khammouane.

At the time when Tha Khọn Yang was being founded the area to the south and on the opposite shore of the Chi—that is, the area lying in what is today Mụang district, Mahasarakham, was nominally within the domain of Rọi-et. Peasants from other parts of Rọi-et territory had begun to settle in this region from sometime early in the 19th-century. For example, oral histories gathered in Tambon Khwao, Mụang district, reveal that several villages in this area of the district which lies nearest to Rọi-et had been settled between about 1815 and 1850.[7] Although there were some settlements in what is Mụang district such as those in Tambon Khwao, the area as a whole appears to have been quite sparsely populated prior to the middle of the 19th-century. According to the chronicle complied by Bunchuai Atthakọn, when the lord of Rọi-et decided to create a new domain in this area in the late 1850's, he sent the future lord of the new domain to find a suitable seat for this domain. At that time, there were no settlements in the vicinity of the place this man chose to be his seat, i.e., the place which was to become the future town of Mahasarakham. (Bunchuai Atthakọn 1962: 48).[8]

In 1859, the Siamese government sent an official to the Northeast to superintend the tattooing of all those who should be on the tax roles. A total of 14,494 men, according to the chronicle of Rọi-et, were tattooed and registered as subjects of the lord of Rọi-et (Phraya Khattiyawongsa 1929: 18).[9] Shortly

7 Bantorn Ondam, one of the researchers who worked in the project under Dr. Toshio Yatsushiro, was told by villagers in Ban Don Du, Tambon Khwao, that their village had been founded 150 years ago as an offshoot of Ban Mi, also located in the same tambon (Bantorn Ondam 1967: 3). This would mean that the village of Ban Don Du would have been founded about 1815–1820 and that the village of Ban Mi would have been founded somewhat earlier. Wasan Tongchai, another researcher in the same project, found that the village of Ban Han in Tambon Khwao had been founded as an offshoot of Ban Don Du. His informants placed the original settlement in the area (at Ban Mi) as having occurred about 200 years previous—i.e., circa 1770—but this date is open to question (Wasan Tongchai 1967: 5). Yet another researcher, Thavorn Koomalayavisai, learned from informants in the village of Ban Iat that their village which is also located in Tambon Khwao, was founded circa 1860 as an offshoot of another previously established village in the area, a village which was most likely Ban Khwao, the present tambon seat (Thavorn Koomalayavisai 1967:3–4). Informants in Ban Nọng Tụn, also in Tambon Khwao, where I carried out my own research, told me of another village, not in the same tambon, which had also been founded as an offshoot of Ban Mi. They also said that Ban Nọng T^{n}n itself had been founded by migrants from Rọi-et "over 100 years ago". It seems clear that Ban Mi was the first settlement in the area, having been founded before 1815. This village in turn gave birth to a number of other villages. Somewhere around 1850, other migrants founded the villages of Ban Khwao and Ban Nọng Tụn and these villages, in turn, expanded.

8 Bunchuai devotes considerable discussion to the question of whether there was more than one village in the vicinity of the present town of Mahasarakham prior to its founding in 1865 and concludes that there was only one.

9 Bunchuai (1962:46–7) gives a figure of 13,000, but Khattiyawongsa's figure is probably

thereafter, the lord of Rọi-et requested the Siamese throne to create a new domain, dependent upon Rọi-et, which would consist of a territory roughly that of Mụang district Mahasarakham today and which would have as its new lord, Thao Mahachai (Kuat), a collateral relative of the lord of Rọi-et. The chronicles speak as though the lord of Rọi-et had decided that his domain had grown too large and that he had a young relative who was deserving of being a lord. Therefore, he recommended the creation of a new domain. In fact, given that there was an old family feud involving the lord and his young relative (Kuat's father had been accused by the present lord of having murdered the previous lord of Rọi-et and had committed suicide to avoid being tried in Bangkok), it is more likely that the initiative for the creation of the new domain actually originated in Bangkok. The Siamese throne would have looked with favor on dividing a domain which had acquired the population as large as that of Rọi-et.

Another factor which may have prompted the creation of a new domain was a drought-caused famine which occurred in 1864–1965 (Mọm Amọrawong Wicit 1963: 73; Phraya Khattiyawongsa 1929: 18; Bunchuai Atthakọn 1962: 50). That such a famine could occur at a time when the population densities were so low compared to what they are today suggests that for at least some parts of the Khorat Plateau, including parts of the domain of Rọi-et, marginal lands had been brought under cultivation. Thus, peasants themselves may have wished to seek better lands.

Be that as it may, the fact is that in 1865, the Siamese court, acting on the recommendation of the lord of Rọi-et, separated 4,000 taxable subjects and a total of 9,000 people from the domain of Rọi-et and placed them under a new domain, that of Mahasarakham (Phraya Khattiyawongsa 1929: 18; Bunchuai Atthakọn 1962: 55–56; Mọm Amọrawong Wicit 1963: 73–4).[10] It seems clear from the context of the chronicles that the founding of Mahasarakham entailed the actual movement of many, if not most, of these 9,000 people from residences elsewhere in the domains of Rọi-et to new homes within what is today Mụang district, Mahasarakham.

The foundation of new settlements in the districts of Kantharawichai and Mụang Mahasarakham as a consequence of the creation of a new domain did not end with the establishment of Mahasarakham. Less than a decade after Mahasarakham was founded in 1865, one Thao Khammun, a young relative of the lord of Mahasarakham, had a falling out with this lord and was able to persuade some 2700 peasants and some of his relatives to move north across the Chi and to resettle near the present town of Kantharawichai. Thao Khammun then persuaded the lord of Kalasin to request that the King create a

more likely to be correct since as lord of Roi-et he would have had access to documentary sources not available to Bunchuai. For further information on this 'census' of 1859, see Mom Amorawong (1963:80).

10 It is somewhat difficult to determine from the chronicles whether the total population is 9,000 or includes 9,000 more than the 4,000 registered persons. I believe that the former reading is correct.

new domain, with Khammun as lord, which would be a dependency of Kalasin. In 1874, King Chulalongkorn complied with the request of the lord of Kalasin and created a new domain known as Kantharawichai (Mǫm Amǫrawong 1968: 87–88).

From the historical reconstruction presented above, it seems clear that the significant population movement into the central Chi River valley occurred following the creation of political domain within this area. This pattern was, in fact, that which led to the populating of most of the Khorat Plateau. Such a pattern does not mean that the peasants themselves played only a passive role in the opening of new lands and founding of settlements associated with these lands. In order to understand the role of both lord and peasant in the founding of settlements in the central Chi River valley, we need now to examine the cultural conceptions which prevailed traditionally in this area regarding rights of access to land which could be cultivated.

3. Traditional Rights of Access to Land

The domains on the Khorat Plateau granted by the Siamese King to Lao (and a few Khmer) lords were defined in territorial terms. However, the lords did not "own" the land within their domains; rather, they ruled those who lived on this land. Rights of ownership, as we shall see in a moment, were embedded in the kinship system of the peasantry. The boundaries of the domains changed through time as the Siamese King created new domains which could be granted to the descendents of established lords or to newly invested lords. The number of domains on the Khorat Plateau increased from fifteen at the end of the 18th-century to nearly one hundred at the end of the 19th-century when the system was abolished.

Lords had the right to collect taxes and to demand corvée service from their subjects. More specifically, they exercised these rights with reference to male freemen who were tattooed and entered upon the rolls of the domain on reaching the age of 20. Once on the rolls of a particular domain, a man remained there for the rest of his life, even if he were to move to another domain, unless he were placed under a new lord by thc act of the King. Étienne Aymonier, who travelled extensively throughout the Khorat Plateau in 1882–1883, reports that:

> When [those subject to corvée] settle in another moeung, even if it be far away, they still retain the lines attaching them to their country of origin and still remain subject to the taxes [of their original Muang]. They regularly receive warnings from the chief of dignitaries or functionaries of their adopted country that immigrants are subject only to political and legal authority, prescribed by the circumstances and public order; they are subject neither to the corvée nor to personal tax (Aymonier 1895 and 1897:I, 179; my translation).

For many of the domains which Aymonier, or his associates, visited they reported that the taxes were collected from so many registered persons living within the domain and so many living outside of it. Most who lived outside were men who had followed the custom of living with their wife's family after marriage.

In addition to being classified according to whether or not they were still residing in the domain in which they had been registered those registered were also classified according to whether or not they had yet married, whether they were "aged" (over the age of 50) or not, and whether they were "deformed" or not. Each class varied in tax liability and in liability for corvée service. Thus, for example, in Mahasarakham province, according to a proclamation (whose date is not given) quoted by Bunchuai Atthakọn, the aged paid a head tax of 2 *salụng* (1/2 baht), the deformed 1 baht, and ordinary registered persons 2 baht. The lords of Mahasarakham also appeared to have taxed migrants who had been entered on the rolls of other domains one baht per year (Bunchuai Atthakọn 1962: 30). In addition to the head tax, ordinary registered persons were subject to periodic corvée mobilization, such as occurred in the 1870's when the Siamese government sent forces to bring an end to attacks in what is today Laos by bandit groups from Yunnan in South China. Finally, all registered persons in Mahasarakham who were married had to pay an annual rice tax of about 2 1/2 kilograms per person (Bunchuai Atthakọn 1962: 30). Aymonier's observations in 14 domains on the Khorat Plateau reveal that the head tax levied against registered persons ranged from 2 1/2 baht per person to 8 baht per person (Aymonier 1895 and 1897). The differences in head taxes appears to have been a function primarily of the amount of tribute which Bangkok demanded of a particular domain.

It should be noted that although head taxes were calculated in money, they were often paid in kind. For example, in Mahasarakham province, many villagers paid their taxes in wild cardamons which were collected in the woods. Six kilograms of wild cardamons was the equivalent of a head tax of two baht in money (Bunchuai Atthakọn 1962: 30). Note also that the rice tax was calculated on an individual basis, not with reference to the size of holdings. It appears to have been the assumption that any domain contained sufficient resources, in both land available for cultivation and in natural resources, for all subjects to be able to meet their own needs as well as the demands for taxes.

With taxes, including taxes levied against agricultural production, determined with reference to those who were entered as registered subjects of the rolls of a lord, it was a matter of indifference to the lord what actual piece of land any particular peasant cultivated. Access to particular pieces of land, then, was determined by peasant custom. In areas of new settlement, homesteading appears to have established the right of a person and his family to particular pieces of property. It is likely that this right must have been necessarily confirmed by successive cultivation of land over the initial few years after homesteading. Once land had been claimed, cleared, and cultivated it became a territorial expression of a specific kin group.

While the rural people of the Khorat Plateau of both the past and present do not have lineages, they do organize themselves into matrilineally-skewed kin groups (Keyes 1975). In this the Lao of northeastern Thailand are similar to the Yuan of northern Thailand for whom a similar matrilineal skewing has also been reported (Turton 1972; Davis 1973). Land is considered to be "owned" by a kin group which at its maximum size in the cycle of expansion and contraction consists of a matrilineal extended family. Ultimate control of the land of such kin groups is vested in the senior couple, or the surviving member of such a couple in the group. The couple or individual exercising such control can divide part of the land and allocate it to one or more daughters and their husbands. On the death of the surviving member of the senior couple, the remaining land will be equally divided among daughters who are married. Given the cultural expectation that after marriage men will reside uxorilocally, at least to begin with, and given the pattern of inheritance, it is apparent that women in northeastern rural society have provided, and still do provide, the continuity in the transmission of land from one generation to the next.

4. Opening of New Lands and Formation of New Settlements in the Central Chi River Valley

From what information is available, it appears that those peasants who founded the initial settlements in the central Chi River Valley never claimed larger amounts of land than could be worked by the labor available in a matrilineal extended family. Given that population has grown steadily by natural increase in the period since the original settlements were founded it is obvious that within one or, at most, two generations, the original holding would be insufficient for those who had inherited it. Those who confronted this situation might be able to add to their holdings by claiming land near to their village which had previously been uncultivated (and, therefore, by peasant custom, unowned). This expansion from the area around an original settlement would lead eventually to founding of satellite villages whose inhabitants moved from the mother village in order to be closer to their fields. The village of Ban Dọn Du, Tambon Khwao, Mụang District, Mahasarakham, was founded in this way:

> Ban Don Du was established 150 years ago as an offshoot of Ban Mee, where villagers lived by rice farming and elephant raising.
>
> Some of them who owned paddy field [sic] far away from the village found it difficult to walk this long distance to their field [sic] and those who owned elephants found it difficult to take the beasts to the marsh, far from the village.
>
> Since the paddy field and marsh were close together, both groups agreed among themselves that they should move out of the village and find a place to live near their work areas.
>
> At last they found the high land, called 'Don' in Thai, where a kind of trees called Pra Du grew in large number. This place was picked out and established as a village which they named Ban Don Du after the high land and the trees (Bantorn Ondam 1967: 3).

The village of Ban Mi (Mee) gave birth to yet other villages in subsequent years. For example, the village of Ban Song Pluai was established in 1895 because "the old village was too crowded for rice farming" (Bamrung Traimontri 1967: 7). Similar patterns of village formation are also reported for other villages in both Mụang and Kantharawichai districts, Mahasarakham (Thavorn Koomalayavisai 1967: 3–4; Mai Chuepaeng 1967: 3; Sawai Pradit 1967: 1–2).[11]

Once expansion out from original settlements had led to the claiming of all suitable land lying between villages, those who faced the prospect of inheriting insufficient land on the dissolution of an estate must needs look beyond the immediate village for land. The quest for land took such people only as far away from their home village as was necessary for them to find uncultivated land. It was rare for only one couple from a village to move out of their home village and to settle on a homestead. More usually, several families would move together and found the nucleus of a new village. Such nuclei, in turn, would attract other homesteaders and a village would then emerge. Such is the case of Ban Kork Ma, Tambon Khok Phra, Kantharawichai District, Mahasarakham:

> Ban Kork Ma was established more than 100 years ago. The first settlers included four or five households under the leadership of Tao Sitthibut... Their exact origin is unknown. Later on, a number of families came from Nong Pan (located in Kalasin Province to the north). Overpopulation and insufficient farm land in Nong Pan brought them to Kork Ma (Wutilert Devakul 1967: 3).

The pattern of response to population pressures on land, then, was for those who lacked sufficient land to look for other land in the Northeast. At the beginning of the 20th century, this pattern became constrained by two factors. In 1901, the Thai government instituted a new land law whereby rights to particular plots were accorded to those whose "ownership" of the plots was attested to in a "title" (*chanot*) issued by the government (Thailand. Land Department, Ministry of Agriculture 1962: 4). While the Thai government to this day has been unable to survey all existing holdings in the Northeast and has, therefore, been unable to issue full titles for all land which has been acquired by inheritance, it has been more successful in restricting access to lands in the public domain to those who obtain legal permission. The Land Law of 1901 did not end homesteading by northeasterners, but its implementation by government officials in the Northeast has made villagers increasingly aware that land is a legally defined commodity. More important than the Land Law in constraining the practice of homesteading was the fact that by the early years of the 20th-century most lands in the Northeast on which people might

11 I found a similar pattern of village formation in Ban Nọng Tụn, Tambon Khwao, Muang District, Mahasarakham, where I carried out research. This community had given birth, at different points in time, to several satellites, but all remained 'neighborhoods' (*khum*) within a single village until the mid-1960's when one of the satellites, Non Khwao Nọi, was officially recognized as a separate village.

homestead were marginal or unsuitable for cultivation given existing methods. Such land as could be cultivated was often land that was periodically—once every other year or every third year—subject to over-flooding, which would result in the drowning of rice plants, or to under-flooding, which would result in the starving of the plants. Again, the availability of only poor land for homesteading did not bring this practice to an end; but it did tend to discourage people from acquiring land unless that was their only viable choice.

Although the population of the Northeast continued to grow, villagers were reluctant to make the choice to go in search of new land which could be found only at places far away from their home villages, which would probably be poor land, and which could be acquired legally only through application to representatives of the government. Rather, people tended to stay in their home villages even when this meant that the population of these villages increased markedly relative to the amount of land available for cultivation. For example, the village of Ban Nọng Tụn where I have carried out research grew from about 200 people at the turn of the century to 800 and perhaps as high as 1,000 at some time in the 1940's. The overpopulation of the village began to place considerable strain on the land available to it. It appears that, like the neighboring village of Ban Iat, the fertility of the soil declined as the amount of labor relative to land increased (Thavorn Koomalayavisai 1967: 4). Under such conditions, some people were forced, whether they wished it or not, to go in search of new land. School records kept in the Ban Nọng Tụn school reveal that between 1935 and 1963, 131 school-aged children "migrated away from the village with their parents". Allowing for the fact that some of these children were of the same parents and that others returned to the village at a later date, there were still at least 75 families which the school records indicate had left the village permanently. There were also a number of families, or individuals, who moved away prior to the time when they had school-aged children.

From about the 1930's, the rice yields in the Northeast declined dramatically both absolutely and relative to units of land and labor (Ingram 1971: 50). It is clear that this decline is directly linked to the increasing population pressures on the land of the Northeast, pressures which were not offset by any changes in the technology of rice production. As Ingram has observed, "the reduced yield suggests that marginal increments of labor and land [were] producing ever smaller marginal physical products" (Ingram 1971: 49). Rice yields continued to decline until the late 1960's when technological changes did begin to occur.

The economic facts of life were not lost on the villagers in the central Chi River valley, or other areas in the Northeast where population pressures were great. In part, villagers attempted to respond to these pressures by seeking land elsewhere even if it meant acquiring land in the Northeast which was poor in quality or going outside of the northeastern region to find suitable land. In the early 1960's, I was told by informants in Ban Nọng Tụn that those who had insufficient land went to such other parts of the Northeast as

Udọn, Kalasin, and Khọnkaen provinces where land—admittedly, marginal land—was still available. In 1974, I was told by people from the same village that some from the community had gone or were considering going to such more distant places as Pracinburi in central Thailand and Phetchabun in north central Thailand. A study carried out in Thoeng (Terng) district, Chiang Rai province, in northeastern Thailand—that is, in a district lying across a rugged mountain chain from the Khorat Plateau—found a significant number of northeastern homesteaders in the district:

> The migrants in Terng began to arrive in bulk about... [1960], coming mostly in March and May of the year. Their method of travel was by bus, often two to four arriving at one time, usually coming from Kalasin, Udorn, Khonkaen, Mahasarakham, Korat, Rọi-et, and Ubol. Two kinds of people arrive: (1) those with money who bought land (at cheaper prices than in the N.E.) and (2) those without land who either squatted or sold their labor to other northeasterners (U.S. Operations Mission to Thailand/Research Division 1966: 3).

While homesteading continues to be undertaken by some northeasterners, it is no longer an adequate response, for reasons given above, to the increasing strain posed by population growth on land resources so long as the traditional form of ecological adaptation is maintained. There is evidence, which we will now consider, that a radical change is beginning to occur in the mode of ecological adaptation, a change which again has involved the active participation of the rural people themselves.

5. Changes in the Mode of Ecological Adaptation

As did their forebearers who first settled in the area, the vast majority of the people living in the central Chi River valley today still devote their main productive energies to cultivating rice. This rice is mainly of the glutinous variety, the variety which forms the staple food of the the peoples of northeastern and northern Thailand and of the peoples of Laos. Land which is suitable for rice cultivation must be low enough to benefit by the annual flooding which occurs throughout the region, but it should not be so low as to permit the rice plants to drown. As the flood waters which come each year have not, until quite recently, been regulated by any significant system of irrigation or flood control works, the floods are not uniform from year to year. In one year, many fields may receive too little water, while in the next other fields may not receive enough. Indeed, by the estimates of most agricultural experts who have studied the region, the majority of the fields in the area receive just the proper amount of flood water in only one out of every three years. The situation is even worse for those marginal fields located on ground too high to receive sufficient flood waters or too low to escape from over-flooding save in exceptional years.

During the dry season of the year, there is inadequate water in all but a small portion of the Chi River valley. Traditionally, the severity of the dry season was ameliorated to some extent by the use of tanks built and maintained by the villagers themselves. But even these tanks, save in rare instances, did not permit double-cropping since the evaporation rate is so high as to dry up the water in most tanks before the new rains come to replenish them.

Given the undulating character of the land of the central Chi River valley (and of most of the Khorat Plateau), a considerable portion of the land is too high to benefit from the annual floods even in the best of years. Moreover, a small portion of land is subject to over-flooding in most years. For these reasons, a significant percentage of the land area of the Chi River valley is not cultivatable at all by traditional means if the product desired is wet rice. In 1963, 40 percent of the land area of Mahasarakham and 45 percent of the land area of Rọi-et province were under cultivation. In the two districts with which we are here concerned, Amphoe Kantharawichai and Amphoe Mụang Mahasarakham, the area under cultivation was 28 percent and 52 percent respectively for the same year (Thailand, Department of Local Administration, Ministry of Interior; U.S. Operations Mission to Thailand; and Thailand. National Statistical Office, Office of the Prime Minister 1965: 23–24).

In the past two decades, the traditional form of ecological adaptation in the Chi River valley has begun to change. Responding to a world demand for fibers, farmers in northeastern Thailand began in the late 1950's to produce increasing amounts of kenaf from which fibers similar to hemp are produced. Land suitable for the cultivation of kenaf is generally only marginally suitable for the cultivation of rice. However, despite the higher income which can be obtained from kenaf rather than rice production, northeastern farmers have been reluctant to convert rice land into kenaf land. Thus, in Rọi-et province, where population pressures are perhaps the greatest in the whole of the Northeast, kenaf production is relatively slight. Mahasarakham province, lying further up the Chi, on the other hand had in 1963 the second largest amount of acreage planted to kenaf of any province in Thailand and Khọnkaen province, which includes part of the upper Chi River valley, had the largest amount of acreage in the country (Silcock 1970: 146–7, 153).

In the past few years, some farmers in the Chi River basin have begun to adopt the new strains of rice and/or have begun the use of chemical fertilizers. The extent to which these innovations have been adopted has yet to be documented. Assuming that petroleum costs do not make the cost of fertilizer prohibitive, it seems likely that the pattern of declining yields which northeasterners have experienced since the 1930's will be reversed. The data for the years between 1965 and 1970 reveal that a reversal has already begun to occur (see Table I).

During the past two decades, the Thai government has instituted programs of planned change which seek to provide better water supply and control for villagers in the northeastern region. In the 1950's, the Thai government undertook, with U.S. support, the construction of large tanks which villagers could use for irrigation throughout the year and which would also permit some local

Table I
Rice Yields, Northeastern Thailand as Compared with Central Thailand and the Whole Kingdom, 1921–1970

Period	Region Northeast	Center	Whole Kingdom
	(piculs/rai)*	(piculs/rai)	(piculs-rai)
1921–1923	4.30	4.24	4.50
1930–1934	3.22	3.91	3.91
1940–1944	2.54	3.37	3.13
1948–1950	2.47	3.90	3.37
1950–1952	2.53	3.70	3.28
1958–1960	2.32	3.62	3.25
1965–1967	2.82	4.65	4.05
1967–1970	3.69	4.95	4.67

Sources: James C. Ingram, *Economic Change in Thailand, 1850–1970* (Stanford: Stanford University Press, 1971), pp. 48–50, 239. Thailand. National Statistical Office, Office of the Prime Minister, *Statistical Yearbook of Thailand* (Bangkok, 1970–1971).

* A *picul* is a measure of weight which equals 60 kg. A *rai* is a measure of area equal to 1,600 sq. meters or 2/5 acre.

flood control. Out of a total of 1,000 projected less than one hundred of these tanks have been constructed. To a great extent, this project gave way to the more ambitious scheme of constructing dams and reservoirs on the Mekong River and its tributaries, a scheme which was developed under the auspices of ECAFE and which was undertaken beginning in the early 1960's. At present (1975), two dams and reservoirs have been built on tributaries of the Chi River —one on the Nam Pong in Khọnkaen province and other on the Lam Pao in Kalasin province. While it is likely that these projects, and others to be built under the Mekong project, will ultimately have a beneficial impact on rural northeastern Thailand, including parts of the central Chi River valley, such an impact has yet to be realized. On the contrary, the projects have thus far been rather disruptive to the economic life of the rural people.[12]

To date, by far the most important change which has had an impact on land use in the Chi River valley has been the growth of a demand for labor in non-agricultural occupations to which significant numbers of villagers from the Chi River valley (and from other parts of northeastern Thailand) have responded. This demand emerged in the early 1950's when Thailand entered into a period of relatively rapid economic growth. Prior to World War II when Thailand was experiencing a slow, but steady rate, of economic growth, most new non-agricultural jobs other than those in the bureaucracy were filled by migrant Chinese. Chinese migration resumed in the immediate post-war years, but was sharply curtailed after Phibun Songkram became Prime Minister again

12 For a discussion of the impact of projects carried out under the auspices of the Mekong scheme, see the following: Demaine and Dixon (1972); Ingersoll (1968 and 1969); Thomas and Block (1970).

in 1947. In 1949 Thailand introduced legal barriers which all but ended Chinese migration into the country. The rail system which had been constructed during the first half of the twentieth century and the new system of roads which was begun after World War II made centers of economic expansion (which consisted primarily, but not exclusively, of Bangkok) more readily accessible to villagers living in more remote areas such as the Northeast. Although considerably more study needs to be made of where northeastern villagers who have responded to this demand for labor have gone, what occupations they have entered into, and how long they have spent in particular occupations, there is no question but that northeasterners have responded in significant numbers to this demand (see Klausner 1972: esp. pp. 91–112; Kirsch 1966; Meinkoth 1962; Textor 1961). In 1963, I discovered that of the 87 men between the ages of twenty and forty who were then living in the village of Ban Nọng Tụn in Mahasarakham province, 62 had worked in one or more than one urban center and 53 of these had worked in Bangkok. To put this in another way, 72 percent of the men between the ages of 20 and 40 had worked in non-agricultural employment outside of their home village for at least a short period of time (one month or longer). The percentage was considerably less for men over the age of 40 and for women: 27 percent of men over 40 and 3 percent of women between 20 and 40 (no women over 40 had been away to work). Most of those from the village who had been away returned to settle down in the village, a pattern which has been reported as occuring elsewhere in Northeast Thailand as well. However, in 1963, six men who had been born in Ban Nọng Tụn were reported as settled in Bangkok or some other urban center. In 1974 I learned that a group of men and women from the village who now considered themselves permanently resettled in Bangkok had formed an association whose main function was the sponsoring of an offering of robes to the monks living in the temple-monastery in Ban Nọng Tụn. In Bangkok I found northeasterners to be the majority of the employees in every service station which I have ever visited, to be employed as the majority of employees in Chinese-managed pottery firms, to be employed in Chinese noodle factories, to be taxi drivers, to be construction workers, etc. Whether temporary or permanent migrants to the city or other centers of economic opportunity, those northeasterners who do leave their villages to take up non-agricultural employment have relieved the demand for land.

In sum, while the patterns of land utilization in the central Chi River valley have only very recently been affected by technological changes in agriculture, they have been significantly influenced since the 1950's by the response of northeastern villagers to the demand for labor in the non-agricultural sector. To put this point in a different way, villagers have come to conceive of labor as no longer restricted to the family enterprises organized primarily for subsistence needs. Indeed, I found villagers in Ban Nọng Tụn well aware of the fact that there was significantly more labor available in the village than could be efficiently utilized in agriculture. This recognition has prompted not only the migration of northeasterners to places where they can take up non-agricultural employment, but it has also begun to stimulate strong interest in family planning.

Unfortunately, there is almost a total lack of data on response of villagers living anywhere in northeastern Thailand to family planning programs now available to them. I can only report some impressions gathered from talks with villagers in Ban Nọng Tụn. In 1963 when I carried out a survey in this village, I discovered that there was a total absence of knowledge about methods of birth control other than abortion. However, this survey also revealed that villagers were definitely interested in learning about such methods. (In fact, shortly after I began my survey, villagers began coming around to consult my wife and myself about what methods they might use.) By the mid-1960's embryonic programs had begun in the Northeast and in the late 1960's, when I again visited Ban Nọng Tụn, I found that most of the villagers to whom I talked were aware that such programs existed. In 1973, I asked several village women whether women in the village were actually using methods of birth control. These several (all of child-bearing age) all replied that "everybody" was using some method (either the loop or sterilization). These replies are unquestionably an overstatement, but they are indicative, I believe, of a strong positive response among northeastern village women to the establishment of family planning programs.

6. Conclusions

The data presented in this paper demonstrate clearly, I believe, that peasants in the central Chi River valley in northeastern Thailand were and are clearly aware of the implications of population growth for their mode of adaptation to the land where they live and make a living. Moreover, it is also clear, I believe, that villagers have taken active roles in responding to such growth. Historically, when local population pressures began to emerge, the peasants responded by either expanding land holdings in the vicinity of their settlements or by seeking new lands elsewhere. In their quest for land, peasants who lived in the central Chi River valley in the 19th-century were only minimally constrained by barriers to access to land imposed by the powerful in society. While it was somewhat difficult to move from one domain to another, there were no restrictions imposed by the lords or by the Siamese king upon peasants who wished to lay claim to previously uncultivated land lying within the domain in which they were born. Moreover, the practice of constantly creating new domains tended to negate any liability which the restriction on moving from one domain to another might have posed.

Legal constraints on acquiring new land were established by the Land Law of 1901. However, by the time this law was implemented in northeastern Thailand, much more important constraints on expansion had begun to emerge. By the 1930's, much of the land suitable for cultivation by traditional means had already been claimed and only marginal land or land totally unsuitable for such cultivation remained. The full import of these constraints was either not realized by the peasantry at first or, as was more likely, the peasants at first knew of no ways to overcome the problems posed by the constraints.

By the 1950's radical changes in the mode of ecological adaptation of the peasantry in the region had begun to take place. Most significant was a re-definition of labor. Labor was no longer seen by the villagers as being for agricultural work alone. It also came to be considered to be available for non-agricultural employment which was increasing in certain centers. Northeasterners, more than rural people from any other part of the country, have taken advantage of non-agricultural employment opportunities. Some changes in agriculture also have begun to occur. Cash crops, such as kenaf, have been added to the production of rice. The new strains of rice and/or the use of fertilizer appear to be being widely adopted. While the promise of the projects undertaken under the Mekong scheme has yet to be realized by the peasantry, I would predict that it will be unless the government mismanages greatly. Finally, it also appears that the peasants are willing to attempt to dampen population growth themselves through adoption of new birth control methods. Again, the success of the family planning programs appears to be less dependent upon the willingness of the peasants to adopt methods of birth control than it does upon the ability of the government and private organizations to make these methods and the knowledge of how they are to be used available to the villagers. For villagers in the central Chi River valley in northeastern Thailand the image of the tradition-bound peasantry which will change only if forced to do so by those who know better—an image which all too often seems to be held by those who plan programs of directed change—is clearly a distorted image.

REFERENCES

Amorawong Wicit, Mom, comp.

1963 "Phongsawadan huamuang monthon isan" ['Chronicle of the Provinces of the Northeastern Circle'] pp. 31–145 in *Prachum phongsawadan phak thi si lae prawat thongthi cangwat mahasarakham* ['Collected Chronicles, Part IV and Local History of Mahasarakham Province']. Mahasarakham: Volume Published for the Cremation of Phra Sarakhammuni.

Aymonier, Étienne

1895 and 1897 *Voyage dans le Laos.* 2 vols. Paris: Leroux.

Bamrung Traimontri

1967 *Summary of Village Attitudes and Conditions in Relation to Rural Security in Northeastern Thailand: An Intensive Resident Study of Song Pluai Village, Tambon Khwao, Amphur Muang, Changwad Mahasarakham.* Bangkok: U.S. Operations Mission to Thailand/ Research Division.

Bantorn Ondam

1967 *Summary of Village Attitudes and Conditions in Relation to Rural Security in Northeastern Thailand: An Intensive Resident Study of Don Du Village, Tambon Khwao, Amphur Muang, Changwad Mahasarakham.* Bangkok: U.S. Operations Mission to Thailand/ Research Division.

Bunchuai Atthakon, comp.

1962 *Prawat haeng phak isan lae mahasarakham bang tǫn* ['Some Aspects of the History of the Northeastern Region and Mahasarakham]. Mahasarakham: Volume Published for the Cremation of Mrs. Pathumma Atthakon.

Damrong Rajanubhab, Prince
1951 "Atthibai chu muang nai monthon udon lae isan" ['Explanation of the Name of Domains in Udon and Isan Circles']. *Warasan Sinlapakon* ['Journal of the Fine Arts Department'] V, 4:20–22.

Davis, Richard
1973 "Muang Matrifocality". *Journal of the Siam Society* LXI:53–62.

Demaine, H., and G. J. Dixon
1972 "Land Tenure Patterns and Agriculture Development in N.E. Thailand: A Case Study of the Lam Pao Irrigation Area in Changwat Kalasin". *Journal of the Siam Society* LX:45–60.

Ingersoll, Jasper
1968 *Human Dimensions of Mekong River Basin Development: A Case Study of the Nam Pong Project, Northeastern Thailand, 1967–1968.* Bangkok: Report Prepared for U.S. Agency for International Development.
1969 *The Social Feasibility of Pa Mong Irrigation: Requirements and Realities.* [Washington, D.C.?]: U.S. Bureau of Reclamation, Department of Interior and U.S. Agency for International Development, Department of State.

Ingram, James C.
1971 *Economic Change in Thailand, 1850–1970.* Stanford: Stanford University Press.

Keyes, Charles F.
1975 "Kin Groups in a Thai-Lao Village". in *Change and Persistence in Thai Society: Homage to Lauriston Sharp*, II. Edited by G. William Skinner and A. Thomas Kirsch. Ithaca, N.Y.: Cornell University Press.

Khattiyawongsa, Praya (Lao na Roi-et)
1929 *Phongsawaden phak isan* ['Chronicle of the Northeastern Region']. Bangkok: Volume Published for the Cremation of Mrs. Si Supha (To Iamsiri).

Kirsch, A. Thomas
1966 "Development and Mobility among the Phu Thai of Northeast Thailand". *Asian Survey* VI, 7:362–369.

Klausner, William J.
1972 *Reflections in a Log Pond.* Bangkok: Suksit Siam.

Mai Chuepaeng
1967 *Summary of Village Attitudes and Conditions in Relation to Rural Security in Northeastern Thailand: An Intensive Resident Study of Song Pluai Village, Tambon Khwao, Amphur Muang, Changwad Mahasarakham.* Bangkok: U.S. Operations Mission to Thailand/Research Division.

Meinkoth, Marian R.
1962 "Migration in Thailand with Particular Reference to the Northeast". *Economic and Business Bulletin* XIV:2–45.

Sawai Pradit
1967 *Summary of Village Attitudes and Conditions in Relation to Rural Security in Northeastern Thailand: An Intensive Resident Study of Nam Thieng Village, Tambon Kok Pra, Amphur Kantharawichai, Changwad Mahasarakham.* Bangkok: U.S. Operations Mission to Thailand/Research Division.

Silcock, T. H.
1970 *The Economic Development of Thai Agriculture.* Ithaca, N.Y.: Cornell University Press.

Textor, Robert B.
1961 *From Peasant to Pedicab Driver.* New Haven: Yale University Southeast Asia Studies, Cultural Report Series No. 9.

Thailand. Department of Local Administration, Ministry of Interior; U.S. Operations Mission to Thailand; and Thailand. National Statistical Office, Office of the Prime Minister.
1965 *Changwat-Amphoe Statistical Directory.* Bangkok: U.S. Operations Mission to Thailand.

Thailand. Land Department, Ministry of Agriculture.
1962 *Anusǫn wan sathapana krom thi din khrop rǫp 60 pi* ['60th Year Commemoration of the Establishment of the Land Department']. Bangkok.
Thailand. National Statistical Office, Office of the Prime Minister.
1970–1971 *Statistical Yearbook of Thailand.* Bangkok.
1973 *1970 Population and Housing Census: Changwat Maha Sarakham.* Bangkok.
Thavorn Koomalayavisai
1967 *Summary of Village Attitudes and Conditions in Relation to Rural Security in Northeastern Thailand: An Intensive Resident Study in Iad Village, Tambon Khwao, Amphur Muang, Changwad Mahasarakham.* Bangkok: U.S. Operations Mission to Thailand/ Research Division.
THOMAS, M. Ladd, and Edward L. BLOCK
1970 *Land Resettlement in Northeast Thailand.* Attendant upon the Projected Pa Mong Dam. [DeKalb, Illinois?]: n.p.
TURTON, Andrew
1972 "Matrilineal Descent Groups and Spirit Cults of the Thai-Yuan in Northern Thailand". *Journal of the Siam Society.* LX:217–246.
U.S. Operations Missions to Thailand/Research Division
1966 *Northeasterners in the Chiengrai Area.* Bangkok: U.S. Operations Mission to Thailand.
VELLA, Walter F.
1957 *Siam under Rama III.* Locust Valley: J. J. Augustin.
Wasan Tongchai
1967 *Summary of Village Attitudes and Conditions in Relation to Rural Security in Northeastern Thailand: An Intensive Resident Study of Han Village, Tambon Khwao, Amphur Muang, Changwad Mahasarakham.* Bangkok: U.S. Operations Mission to Thailand/ Research Division.
WUTILERT DEVAKUL, M. R.
1967 *Summary of Village Attitudes and Conditions in Relation to Rural Security in Northeast Thailand: An Intensive Resident Study of Kork Ma Village, Tambon Kok Pra, Amphur Kantharawichai, Changwad Mahasarakham.* Bangkok: U.S. Operations Mission to Thailand/Research Division.

Contributions to Asian Studies, Vol. IX

Land Use and Population Growth in Colonial Ceylon[1]

PATRICK PEEBLES

University of Missouri, Kansas City, U.S.A.

THE CHANGES in population and land use in Ceylon have been of such revolutionary proportions in recent decades that one might lose sight of the extent of change in these areas in the nineteenth century. This paper examines some aspects of the macro-level changes in population and analyzes the means by which new land was made accessible for cultivation. For most of the century increases in cultivated area outpaced population growth, but most of the increase came in plantation crops and coconuts rather than wet rice, and there may have been a substantial decline in shifting cultivation. Some implications of these developments are suggested in the last section of the paper.

The first section of the paper considers the exogenous factors in Ceylon's agriculture, particularly innovations resulting from British colonial policy. I stress the importance of crown land administration, which is the subject of the second section. Finally, I attempt to provide better estimates of Ceylon's population growth during the century and to relate population pressure to other broad aspects of Ceylonese economic organization.

Exogenous factors

One model of change in agriculture in nineteenth century Ceylon identifies the plantation industry, irrigation and grain taxes as the major agents of change in the nineteenth century economy (Ameer Ali 1970: 266). The first modernized the economy, the second made peasant agriculture productive and the third forced peasants into a money economy. These resulted in a disequilibrium of land and agrarian population, according to the model, because British crown land policy blocked access to new land. Not only was the price of crown land too high, but the administration of crown lands was so understaffed at the top and dominated by corrupt Ceylonese officials at the bottom that it could not respond to peasant demand (Ameer Ali 1970: 268–279). This argument oversimplifies the extent of changes in the rural economy in general, and the effects of crown land administration in particular.

1 The research on which this paper is based was made possible by a grant from the Foreign Area Fellowship Program of the American Council of Learned Societies and Social Science Research Council. The data processing was assisted by a Dissertation Completion Fellowship from the National Science Foundation and by support from the History Department of the University of Chicago.

The role of the British colonial government must be analyzed in terms of its actual accomplishments and their consequences rather than its policy decisions and their logical implication. Irrigation, for example, was a preoccupation of government spokesmen as a contribution to peasant welfare, but little action was taken until 1871, except for a brief flurry of interest during Henry Ward's term as governor, 1855–1860 (University of Ceylon 1973: 153–156). Even in the last three decades of the century irrigation expenditures were irregular and their greatest impact limited to a few major irrigation works. At the other extreme, the alienation of crown lands was not emphasized as an aspect of government policy, but proved to be one of the most pervasive agents of change in the nineteenth century. The government "granted" 1.6 million acres of land in the century, less than a quarter to European planters. The remainder became the freehold property of Ceylonese villagers, entrepreneurs, and monastic institutions.

The difference in importance between the crown grant system and irrigation was economics: whereas irrigation works did not pay for themselves, the sale of crown land returned substantial amounts of revenue to the colonial treasury. In theory the principle of economic viability and the principle of promoting the general welfare were harmonious, but in practice the second was always subordinate to the first. Noble ambitions often fell victim to recurring budget deficits. For example, the British encouraged private enterprise from the assumption of crown rule in 1801, but the need for revenue forced them to hold on to many aspects of the mercantile system—most notably the cinnamon monopoly—until 1833 (University of Ceylon 1973: 48).

The most effective British endeavours were those that cost little and provided immediate returns, such as road construction. The island's outstanding network of roads was built at first with *corvée* labor and then with a road tax. A system of tolls provided a substantial portion of government revenue. Railroad construction was delayed until 1867 by the Colonial Office's refusal to guarantee its loans, but its financial success soon resulted in the spread of lines from Colombo to all the populated parts of the island (University of Ceylon 1973: 308–309).

It is not surprising, therefore, that plantation investment was eagerly sought. The plantation industry became the dominant sector of Ceylon's economy and the primary focus of British administration, and will be treated in some detail. A favorable crown land policy was instrumental in the growth of the coffee industry (Peebles 1973: 132–135). The reforms of a commission of enquiry promoted private investment in land from abroad by the easy access to land in fee simple, free of taxes. After July, 1833, crown land was auctioned at a nominal price. The coffee boom raised the question, somewhat belatedly, of what exactly was crown land. Two ordinances (No. 12 of 1840 and No. 9 of 1841) declared "that all Forest, Waste, unoccupied and uncultivated Lands, shall be presumed to be the Property of the Crown, until the contrary thereof be proved," including *chenas* (shifting cultivation).

The administration of crown land sales in these years is a sordid episode in British colonial history. The civil servants who sold the land were themselves large speculators in coffee lands. While the crown lands ordinances were being passed they "deliberately and consistently misled the Colonial Office" in order to expropriate as much land as possible (de Silva 1964: 39). The Surveyor-General and the Government Agents (abbreviated G.A.) bickered over which land to survey and sell. Civil servants neglected their duties to attend to their own estates. Until 1844 the initiative was left to the planters. The Surveyor-General could not keep up with the demand and by 1843, there were 186 applicants for 44,000 acres outstanding in the Central Province alone. As a result informal agreements among planters and between the planters and the G.A.'s circumvented the authorized procedures.

The application for a piece of land came to be accepted as being as good as a title deed. According to the Surveyor-General, speculators applied for lands with no intention of developing them. These lands were then reserved for the applicant to purchase when convenient with no obligation to purchase them. Lands sometimes were sold without surveys. Speculation was unchecked until 1844 when the upset price was raised to 20 shillings and the sale of land without a survey was prohibited. Approximately 300,000 acres of crown land were sold as coffee lands by 1847, when financial difficulties in England shut off the flow of capital.

Estate development proceeded alongside the land speculation. It was estimated that 52,722 acres of coffee estates were under cultivation in 1848 on 367 estates, most of them near Kandy in the central province. Altogether there were no more than 500 European planters and merchants at this time. About 35 were agents for the 23 small trading firms in Colombo and the rest actively engaged in planting. Several Ceylonese investors bought coffee land in the early coffee boom, but only one, Warusahännädige Joronis de Soysa, invested at the same scale as the Europeans.

The development of the plantation economy was cyclical. The first crisis came in 1847–48, when the first large-scale plantings came into full bearing at the same time that new investment stopped. In 1848 the production was "greater than the whole consumption of Great Britain," and prices fell one-third. The industry revived in the 1850's and the area under cultivation reached 200,000 acres by 1860. Through the 1860's and 1870's coffee dominated Ceylon's economy. It reached a peak of over 300,000 acres in 1877 when the coffee-leaf disease (*hemileia vastatrix*) began to destroy trees faster than they could be replanted. The emergence of the Brazilian coffee industry put the final touches on the demise of the coffee plantations. The tea industry rapidly replaced coffee, so that the plantation structure remained.

The presumed effects of the plantation economy on nineteenth century Ceylon have been neatly listed as:

> Increasing monetization, increasing commercialization, the growth of bazaar towns and other urban and semi-urban settlements, the proletarianization of some segments of the indigenous population, the widening of opportunities for the profit motive and production

> for exchange, the development of a land and share market and increased sales of land, new legal concepts, demographic shifts, population pressure and localised regional scarcities of arable land, the contraction of swidden agriculture in the Wet Zone, an addition to the plurality of a plural society, and the development of a dualistic form of economy (University of Ceylon 1973: 117–118).

The view that the dual nature of Ceylon's economy prevented any "leakage" between the two sectors has been effectively challenged (University of Ceylon 1973: 162–164), but no consensus has been reached on the total impact. Three suggestions have particular relevance for this paper and will be re-introduced later: 1) the coffee and tea estates provided cash incomes for laborers, small-holders of coffee, carters, suppliers and carpenters, and this cash income was eventually invested in land; 2) the coconut plantation and graphite industries used the export facilities of the coffee plantations and were largely products of local capital, labor and entrepreneurship; 3) the means by which crown lands were made accessible to foreign investors were also available to Ceylonese investors. In these three areas, at least, the plantation economy deeply influenced the domestic economy.

To some extent the influences attributed to the plantation economy were the effects of British attempts to establish proprietary rights in land and to extend "native" cultivation. These were policies from the time of the first governor, Lord Frederick North. Land in early nineteenth century Ceylon was differentiated in regard to its alienability and the amount of produce accruing to the cultivator (de Silva 1962: 322–331; Bertolacci 1817: 283–288, 318–321). At one extreme there was *paraveṇi* (heritable) land which resembled free-hold in many respects. At the other were government estates cultivated by hereditary tenants or employees of the government. The government's claims to land or to a share of the produce took several forms. *Paraveṇi* was subject to a tax of a fixed proportion of the produce that varied widely according to the fertility of the soil, water supply, and means of cultivation. However, only the tax on paddy lands was collected; the Dutch attempted but failed to enforce a small tax on coconuts, *jak* trees and other "high land" produce and the British were no more successful. Rights to *paraveṇi* were freely transferable in theory, not only by inheritance (*urumē*) and purchase (*milaṭa*), but by gift (*tägi*) or prescription (*buttiya*). The "rule" for inheritance was equal divisions among heirs, but testaments were known and children could be disinherited. Virtually all *paraveṇi* prior to the eighteenth century consisted of village paddy fields and high-lands. New fields were cleared if the village fields became crowded: "everyone had a sufficiency of land for the maintenance of himself and his family, and there was always an abundance of forest available" (Pieris 1920: 26–27).

The villager's *paraveṇi* land was a proportion of the village field, a precisely determined fraction that was readily transferrable (Obeyesekere 1967: 17–21). However, it was this share (*paṅgu*) and not a fixed plot of land that was owned. The field was partitioned into sections, which would rotate among the share-holders to ensure an equitable distribution of good soil and irrigation. The villager's claim to the forests in the vicinity of the village was analogous to his *paraveṇi* tenure. He did not claim the forest or any particular section of it, but the right to practice shifting cultivation (*chena*) or to clear an area for permanent cultivation (*kanuhis paraveṇi*).

The Portuguese and Dutch perpetuated a variety of tenures that they deemed to be direct ownership by the sovereign, with some modifications (Bertolacci 1817: 277–288; Codrington 1937: 59–63). These were cultivated by means of a variety of tenures. The *muttețțu* was a section of the paddy lands in some villages that villagers were required to cultivate as part of their compulsory service to the king. It and other land that paid revenue into the royal treasury was called *gabaḍagam*. *Ratninda* was the "king's exclusive possession"; *malapalā*, grants resumed for failure of heirs; and *nilapalā*, grants resumed for nonperformance of service. One crucial category, *ratmahara*, is still unclear. The British called it "the waste in general," but the definition is self-serving, since it was this land that had already been appropriated as "waste land" and was being sold as crown property. The idea of "private title" was foreign to the system, with the exception of land grants given as a royal reward.

The Dutch East India Company discouraged the accumulation of property by Ceylonese because it interfered with the Company's mercantile enterprises. In spite of this attitude the end of the Dutch period and the early British period were years in which many Ceylonese acquired title to a great deal of land. The Dutch granted and sold land in the late eighteenth century to provide revenue and to increase production of cash crops. In the confusion of the Anglo-Dutch wars the grantees claimed these and other conditional tenures as private property. At the same time many government-owned gardens scattered throughout the Colombo district were taken over by Ceylonese.

For the British the tendency towards increased private ownership in Ceylon was a hopeful sign, the key to a viable colony. Governor North's commission of March 13, 1801, ordered him to resume land grants given to Europeans and to "observe a liberal system of Policy towards such of the Natives as may apply to you for grants of land," in order to foster the development of agriculture.

As always, however, policies yielded to financial stringency. North's successor, Thomas Maitland reversed some of North's most ambitious programs (de Silva 1962: 550). He abolished the survey department to save money, reinstated service tenures, and continued to enforce *corvée* labor (*rājakāriya*) for which North had substituted a tax on rice cultivation. (The tax was still collected). Maitland, however, did grant many of the Dutch gardens to Ceylonese cultivators. In the course of the nineteenth century additional measures were taken to establish the rights and obligations of private property. These included the licensing and discouragement of shifting cultivation, registration of monastic landholdings, recognition of prescriptive rights against the crown, the law of partition, the commutation of grain taxes for cash payments, and an abortive attempt to register title to land. Further research is necessary before the impact of any of these can be measured with any degree of confidence. Some aspects of their implementation can be suggested, however.

The enforcement of proprietary rights in the courts put teeth into legislation. One could lose or acquire property through litigation in any of these ways; land speculation by means of a civil suit seemed to become a national pastime. The law of partition will serve as an example (Obeyesekere 1967: 179–184). Under the Roman-Dutch law of partition enforced by British courts, a shareholder in a paddy field was entitled to compel the division or sale of the field in order to possess his share (Wickremesinghe 1969: 3–7). In 1864 an ordinance to facilitate such a division or sale was put into operation. Its purpose was to destroy the normal system of undivided co-ownership of wet-rice fields, which the British believed, incorrectly, to tend toward the uneconomic fragmentation of shares. To end it the partition ordinance empowered courts to grant indefeasible title to the partitioned land (Wickremesinghe 1969: 369–370). Within a few years it was believed that large amounts of wet-rice fields were getting into the hands of speculators who would purchase a share and institute partition proceedings. Co-owners who owned shares too small to cultivate would be forced to sell their shares or to agree to the sale of the whole field. The official records on land transfers due to partition actions are misleading, since the very threat of a suit may have been sufficient to force sales that would never be recorded.

Prescriptive rights evolved by judicial decision (Peebles 1973: 223–224). Governor North's Proclamation of March 1, 1801, declared that "undisturbed possession of lands... for 30 years shall be equivalent to and constitute a prescriptive right..." to those lands. In 1803 the prescriptive term was reduced to ten years. These proclamations were repealed by Regulation No. 13 of 1822, then reinstated by Ordinance No. 8 of 1834. All legislation on the matter refer specifically to "prescription of actions" rather than to the Roman *usucapio*, the acquisition of ownership by prescription. In spite of this, the courts of Ceylon recognized *usucapio* by 1870 at the latest. More important, the administration began to recognize the prescriptive rights of encroachers. The government empowered Agents to issue "certificates of quiet possession" to the holders of land that had been cultivated for ten years or more and to "sell" land recently brought into cultivation at a price of one-half of the improved value. These mitigated the harshness of the Crown Lands Ordinance. They demonstrate the process of adaptation to the two systems: a cultivator was still allowed to bring new lands into cultivating (although this was still technically illegal), but he was required to obtain title for that land from the Agent.

Shifting cultivation was a much more difficult problem for the government of Ceylon (Peebles 1973: 224–225). Whereas ordinary "encroachments" could be viewed with equanimity, and even approval, because they extended cultivation, *chena* was believed to be a primitive and inefficient form of cultivation that ultimately destroyed the fertility of the land. Under the British scheme they fell clearly into the category of crown lands because the land was not cultivated more than once every seven or eight years or more and because no title had ever been given to *chena* land. On the other hand, some *chena* was always under cultivation, as the villagers cleared a new site each season. It was an integral part of village life, and its loss would create hardships for many villages. The British were unwilling either to yield on their claims to all land not cultivated regularly or to attempt to prevent shifting cultivation entirely. To extricate themselves from the dilemma, they invented a distinction between "private *chena*" and "forest." "Private *chena*" was that forest land on which *chenas* had grown previously; the Crown Lands Ordinance protected such land in the Kandyan districts. Shifting cultivation was prohibited in the coastal regions and in all virgin forest. Where allowed, it was licensed.

The total effect of these measures seem to have achieved the end desired by the British, the creation of proprietary rights in peasant land. The evictions from encroachments, the limited licensing of *chena*, the grain tax expropriations, and the sales through partition suits all suggest a rapid turnover in land-holdings. The effect of these, whatever further research may show them to have been, was probably mitigated to some extent by the increase in cultivated area. Much of this increase was on land alienated by the crown, to which I turn now.

Crown Land sales

The predecessor of the system of crown grants was the privilege of the governor to issue free grants of land at his own discretion. Before this privilege was withdrawn in 1833 over 50,000 acres of land were granted in freehold (*sinnakara*) to Ceylonese cultivators (Peebles 1973: 98–99). Most of this land had been Dutch "gardens" that had been planted with spices in the later years

Table I
Selected Total Grants

Grantee	*No. of Grants*	*Acreage*
Solomon Dias Bandaranaike	12	501.1
David de Alwis	5	286.1
Johannes Paulus Perera	5	264.2
Marcellis Perera	6	212.0
Christoffel de Saram	1	104.0
Abraham de Alwis	7	103.4
Hendrick Dassenaike	3	88.6
Johannes Louis Pieris	6	87.8
Celestinu Fonseka	4	73.9
Andris Amarasekere	2	67.9
Simon de Livera	6	64.6
Matthew Gomes	2	63.7
John Leonard Philipsz Panditaratne	3	63.1
Peter Harmanis Perera	1	56.7
Abraham de Saram	2	56.2
Hendrick Alphonso	4	55.8
Samuel Ameresekere	5	52.0

Source: Ceylon National Archives, File 6.

of the eighteenth century. The British were attempting to stimulate agricultural development by breaking up these gardens—some of which covered hundreds of acres—into small plots of land and issuing title to the plots to villagers. These grants tended to concentrate in relatively few hands.

The summaries given in Table I have been laboriously compiled from records of the Ceylon National Archives and are certainly incomplete. Even so, they show that at least a score or so Ceylonese were granted substantial amounts of land—50 acres or more. These men were landed proprietors on a scale not known in Ceylon earlier, the forerunners of many such men to emerge in the nineteenth century.

The British, trying to promote cultivation, made ability to cultivate the sole criterion for land grants, and these grantees convinced the British that they could develop the land granted to them. The *mudaliyars*, traditional local officials of high family status, claimed that agricultural laborers of various castes would work for them in return for food and entertainment. Unfriendly witnesses testified to the Colebrooke-Cameron Commission, however, that such laborers were impressed in the name of *rājakāriya* (corvee). All of the grantees are *mudaliyars* and all but three are of the *goyigama* (highest) caste.

These land grants created a category of landed proprietors with landholdings of unprecendented extent (Peebles 1973: 100–105). Prior to this there were few landed proprietors of any magnitude. In 1834 Governor Horton provided a list of the leading *goyigama* and *karāva* jurors who were eligible for the post of Sinhalese representative on the newly-created legislative council: Thirty of the 91 "first-class" *goyigamas* and 12 of the 34 "first-class" *karāvas* are described as owners of fewer than 25 acres, some with five or fewer acres. One can infer that fewer than 100 household heads (as they appear to have been) owned land in extents comparable to the amounts granted.

Proposals in the areas of land and agriculture were central to the Colebrooke-Cameron reforms of 1833 (Peebles 1973: 125–126). Colebrooke believed that economic development should be in the hands of private investors and proposed reforms, such as the abolition of *rājakāriya*, that would promote enterprise. He recommended a general survey of the island in preparation for the sale of crown land.

Table II summarizes the sale of all crown land from 1833 to 1889. This comprised nearly ten percent of the surface area of the island. This figure is inflated by the inclusion of all types of land to which the government recognized title, such as temple land (*ca.* 360,000 acres) and peasant smallholdings to which title was recognized. The plantations in which Europeans invested never covered more than 450,000 acres during the nineteenth century and some of this was owned by Ceylonese planters. In Ceylon much more land was granted in freehold to Ceylonese cultivators and planters than to Europeans, even excluding those "grants" that represented no change in landholder.

There were three distinct periods in the alienation of crown lands: 1833–44, 1845–58, and 1859–89. During the first of these periods—the "coffee mania" when a small number of British civil servants and merchants resident in the

Table II
Crown Grants, 1833–1889

Period	*Lots*	*Lands Sold and Settled* *Acreage*	*Average*
1833–1844	2,764	275,078	99.5
1845–1849	879	31,095	35.4
1850–1854	401	16,158	40.3
1855–1859	3,606	77,936	21.7
1860–1864	18,468	149,403	8.1
1865–1869	32,599	195,442	6.0
1870–1874	27,819	216,049	7.8
1875–1879	21,178	184,208	8.7
1880–1884	26,660	127.925	4.8
1885–1889	17,555	94,402	5.4
1833–1859	7,650	400,265	52.3
1860–1889	144,279	967,429	6.7
1833–1889	151,929	1,367,694	9.0

Source: *Crown Grant Abstracts.*

island acquired huge tracts of land—few Ceylonese had any opportunity to compete. The average size of the lots is much smaller than the average coffee plantation because there were over 2,000 lots sold or granted free to smallholders at the same time. In the Western Province in this period 2,158 lots under 100 acres were sold covering 13,695 acres according to the *Blue Books.* Although these figures may not correspond to the Survey Department figures, it gives a better approximation, 431 acres for the size of the average coffee estate. Crown land sales did not revive after the rise in the sale price and the crisis of 1846–47 until the late 1850's. As late as 1857 only 80,950 acres of coffee were cultivated, much of it "resuscitated" from abandoned estates (Mills 1964: 237). The sustained growth of the coffee industry appears to have stimulated further investment in land, and nearly a million acres of crown land were sold in the 31 years from 1859 to 1889. The large number of small grants show that many new investors were purchasing land.

Approximately 650,000 acres of this land was waste land sold at auction by G.A.'s.[2] At least 100,000 acres were lands for which "certificates of quiet possession" were issued or which was sold at one-half improved value. These latter usually were lands already in cultivation for an undetermined length of time at the time of alienation. The remainder apparently consisted of lands

2 The land records used in this paper are described in detail in my dissertation, Chapter VII. They are drawn from the records of crown grants in the Ceylon National Archives and are an attempt to overcome the absence of standard landholding records such as those used in nineteenth century India. The samples are described in the appendix (1973:352–357).

settled by all other means, including monastic landholdings, business lots, and those disposed by judicial decisions. The figure is also inflated to an unknown extent by transactions not completed after the sale was recorded.

A one per cent random sample of the 64,341 deeds of sale was taken for the period 1860–1889. Tables III and IV summarize the results. Only about 4,800 lots were purchased by Europeans, but they contained the largest grants. The data do not allow a reliable estimate of the total land purchased by the Europeans, but the total could not have been more than 250,000. This means that Ceylonese investors acquired 400,000 acres of crown land in addition to the 100,000 acres of smallholdings to which title was given, any acquisitions before 1860, and all private purchases from Europeans. The system of crown grants, therefore, made the ownership of substantial amounts of land available to Ceylonese.

Table III
Distribution of Crown Grants 1860–1889

Race	*Percentage*	*Date*	*Percentage*
Sinhalese	68.2	1860–1864	10.2
English	6.7	1865–1869	13.3
Burgher	0.6	1870–1874	23.8
Tamil	14.7	1875–1879	17.8
Muslim	8.9	1880–1884	17.7
Companies	0.9	1885–1889	17.2

Source: one per cent random sample

Table IV
Size of Crown Grants to Ceylonese and Europeans

Price (Rupees)	*Ceylonese Percent*	*European Percent*	*Total Percent*
0–9	23.7	6.8	22.6
10–100	55.2	27.3	53.4
101–500	16.7	34.1	17.8
501 and over	4.5	31.8	6.3
Total	100.0	100.0	100.0
Acreage	*Ceylonese Percent*	*European Percent*	*Total Percent*
0.0–0.9	37.3	18.2	36.0
1.0–4.9	39.4	18.2	38.0
5.0–9.9	10.6	4.5	10.2
10.0–49.9	10.1	34.1	11.8
50 and over	2.5	25.0	4.0

Source: one percent random sample

It is evident that the crown grants enabled wealthier investors to accumulate sizable holdings. From the distribution of grants alone it is clear that at least some Ceylonese bought large estates. Some of these lands were in Kurunegala and Colombo Districts, where the average price of crown land was substantially higher for Ceylonese than other land (Table V). This indicates the interest of Sinhalese planters in the coconut lands in those districts.

Table V
Average Price of Grant by Race and Region

Race	*Region*	*Number of Deeds*	*Average Price (Rupees)*
European		43 (6.7%)	1315.9
	Central Province	25 (3.8)	1677.8
	Other	18 (2.8)	813.2
Ceylonese		610 (93.3%)	140.2
	Kurunegala District	35 (25.4)	291.1
	Colombo District	159 (24.3)	198.3
	Eastern Province	87 (13.3)	156.1
	Northern Province	24 (3.7)	57.0
	Other	305 (46.7%)	95.2
Total		653	217.7

Source: one per cent random sample

A long and tedious analysis of two samples of crown grants was undertaken in order to estimate the extent of investment in crown lands by the colonial elite and to identify the largest investors (Peebles 1973: 245–252). The two samples consisted of all deeds that had a sale price of more than £50 (before 1872) and Rs. 500 (after 1872), and all deeds issued to prominent Sinhalese purchasers. The latter were selected from a long list of individual and family names identified in other sources.

The four per cent of all deeds sold for more than £50 or Rs. 500 accounted for about forty per cent of all lands sold—247,500 acres. Two-thirds of this land was purchased by Europeans, but there were 1350 deeds totalling 83,700 acres purchased directly by Ceylonese investors. These purchases represent only

Table VI
Distribution of Land Grants Over Rs. 500

Lots	*Group*	*Mean Acres*	*Total Acreage*
1085	Europeans	150.5	163,800
293	Moratuwa *karāvas*	108.9	31,800
108	*Goyigama* mudaliyars	80.0	8,600
142	Other Prominent Sinhalese	71.5	10,200
369	Other Sinhalese	31.2	11,500
439	Other Ceylonese	49.4	21,600
2432	Total	101.8	247,500

one aspect of the economic activities of Ceylonese elites, but they provide an index to the entrepreneurial activities of the colonial elites and aspirants to elite status. They in fact underestimate the total purchases of waste land by these individuals, since the individuals who bought this land also bought many —in two cases hundreds—of smaller grants. The results are given in Table VI.

The 2,432 deeds were also sorted by their potential land use. The dry zone and wet zone were divided, and the latter separated into the up-country coffee and tea plantation land and the low-country coconut and cinnamon land. The special cases of urban property in Colombo and lands in which graphite was mined were also identified. All but 20 per cent of the deeds were classified in this manner. This residual category includes much land that could be either coconut or coffee land, plus residential or commercial property, and lands that could have been used for other groups. Table VII cross tabulates the four ethnic categories and the land uses. Each group had a different pattern of land purchases. Europeans, of course, bought virtually all of the coffee land. Sinhalese investors preferred coconut land. Dry zone land, much of it in the rapidly-developing east coast near Batticaloa, was purchased by Muslims and Tamils.

Table VII
Waste Land Use and Ethnic Group, 1860–1889

Land Use	*Prominent Sinhalese*	*Other Sinhalese*	*Other Ceylonese*	*Europeans*	*Total*
Colombo town[1]	13	4	4	7	28
Coconut land[2]	366	194	126	43	729
Coffee Estates[3]	77	30	21	745	873
Graphite mining[4]	37	35	5	1	78
Dry Zone[5]	2	7	212	27	248
Unclassified Wet Zone[6]	45	99	71	261	476
Totals	540	369	439	1084	2432

Includes all deeds of land sold for Rs 500 or more, 1860–1889.
1 Includes only those lots within the municipality sold as waste land. Most urban property was sold separately as "building lots."
2 Land in districts in which the cultivation of coconuts increased greatly in the period.
3 Lots over 50 acres sold for more than Rs 15 per acre in coffee planting districts.
4 Small lots sold for over Rs 100 per acre in certain localities.
5 Districts with less than 75″ annual rainfall, excluding coconut and coffee land.
6 A residual category.

The crown land sales therefore were unequally distributed among the Ceylonese population. A few prominent Ceylonese families accumulated enormous estates. Further analysis identified two sets of intermarried families, one *karāva* caste and one *goyigama* caste, who purchased a disproportionate share of crown land purchased by Sinhalese elites (Peebles 1973: 254–257). Even at the village level the crown land must have contributed to inequality, since only a minority of cultivators could have purchased any land at all. For the implications of this, we must examine population growth and its pressure on the cultivated area in the nineteenth century.

Population growth

There are no accurate figures for the population of Ceylon at the beginning of the nineteenth century, but aggregate totals can be estimated. The University of Ceylon *History of Ceylon* asserts that the population of Ceylon then was one million or less, "making every allowance for underestimation" (1973: 146). This figure implies a higher growth rate—1.4 per cent per year—for the first half of British rule in Ceylon than for the second half. It is based on the imprecise results of early British census and appears to be implausibly low. A better estimate can be derived by using Sarkar's estimates of the population in 1871 and working backwards (1957: 22).

Sarkar's estimates of the population of Ceylon before 1871 rest on several tenuous assumptions: that the degree of under enumeration was constant from one estimate to the next, that the estimates for the newly-conquered Kandyan districts were no more erroneous than those for the coastal districts, and that population growth rates for the coastal region can be applied to the interior (Table VIII). His attempt to combine unreliable contemporary estimates has given some implausible results, such as a 4.408 per cent average annual decline from 1809 to 1814. This is explained by the famine conditions of 1811–1813. However, 1799, 1805, 1823, 1824 and 1825, were also famine years, but no account has been taken of them (de Silva 1962: 354–355). In fact the period 1814–1827 shows a rapid and unexpected rise in the population, which should be attributed to decreasing underenumeration between the first and second British census rather than population growth.

My revised estimates for the period 1789 to 1871 are given in Table VIII. I have used Sarkar's estimates of the rates for 1837 to 1847 and 1847 to 1857, based on the figures in the *Blue Books*. For the period 1857 to 1871 he estimated the rate as the average of the rates for the uncorrected populations in the preceding and following decades. I have averaged the corrected rates. From 1789 to 1827 Ceylon passed through a series of crises that would have tended to

Table VIII
Estimated Population of Ceylon

Year	*Sarkar (000)*	*Annual Rate*	*Revised (000)*	*Annual Rate*
1789	1788	—	1602	—
1809	1533	−0.717	1602	0.0
1827	1445	−0.300	1602	0.0
1837	1734	+4.112	1806	+1.195
1847	2148	+2.389	2287	+2.389
1857	2385	+1.105	2552	+1.105
1871	2820	+1.303	2820	+0.950
1881	3044	+0.794	3044	+0.794
1891	3330	+0.940	3330	+0.940
1901	4031	+2.105	4031	+2.105

Source: Sarkar 1957:22.

decrease the population: the financial collapse of the Dutch East India Company, British conquest, a rebellion in Colombo district, British and Kandyan warfare, the Kandyan rebellion, and the famines mentioned above. I have left the growth rate at zero, representing a fluctuating rather than a stable population.

How did so many people elude the early census-takers? It was assumed that rice was the staple item of consumption and that most rice was produced in wet fields. Shifting cultivation was believed only to have provided a supplement to make up deficiencies (de Silva II, 354–356). Since a demand existed for imported rice, the early nineteenth century observers concluded that the population could not have exceeded by much the numbers fed by domestic wet-rice production and imports. They estimated that the Ceylonese each consumed six *parras* of rice or 250 to 260 pounds each year. As James Brow points out in this volume, however, there are decisive economic advantages to shifting cultivation and evidence that it was widespread in the nineteenth century.

Large areas of the wet zone were used for shifting cultivation in the early nineteenth century. Estimates of the extent of shifting cultivation ranged up to a half of the wet-field rice, but it is not clear on what basis the comparison was made (Ameer Ali 1970: 41). In any event, the British estimates are bound to be underestimates, as their low regard for shifting cultivation, the history of sanctions against it and the cultivators' awareness of these facts encouraged the villages to conceal their *chenas*.

There was a vast increase in the cultivated area of Ceylon in the nineteenth century. Most of this increase came in plantation crops, however than peasant agriculture. There is no reliable way at present to estimate population pressure before 1871, particularly since so little is known about shifting cultivation. Published estimates of the peasant population per acre of cultivated peasant land show a relatively stable ratio for the late nineteenth century (University of Ceylon 1973: 153; Snodgrass 1966: 48–49). These estimates omit reference to *chena* land, however, and it is generally assumed that the extent of *chena* was decreasing.

The rapid increase of coconut cultivation undoubtedly absorbed much of the actual or potential *chena* land of the Western, Northwestern, and Southern Provinces. The areas of the most intensive coconut cultivation are the only areas of the low-country wet zone where shifting cultivation was not still common in the mid-twentieth century (cf. Maps in Farmer 1950: 56). In the nineteenth century the common means of clearing the land for a Ceylonese coconut plantation was to lease it out as a *chena* for the first year, as the British observers mention with distaste.

Even if the overall ratio of peasant to land was not increasing, it is likely that the distribution of peasant land was becoming increasingly unequal (Peebles 1973: 259–267). The statistics on crown land sales support this, and all indications point to a crisis in the 1880's that resulted in the large scale sales of paddy land by villagers. There are five explanations related to the government policy of establishing proprietary rights given for this phenomenon:

1) the collapse of the coffee industry shut off the cash income of many villagers; 2) the Partition Ordinance of 1863 enabled one shareholder to force the partition (and thereby the sale of uneconomic units) of the entire field; 3) the continued availability of coconut land, which was untaxed, made it a more appealing investment for villagers as well as elites; 4) the Grain Tax Ordinance of 1878 enforced a "compulsory commutation"—payment in cash on either an annual basis or as a proportion of the value of the crop—in many of the Sinhalese districts; and, 5) the Service Tenures Ordinance of 1870 made it possible for purchasers of land held by tenants of temples to do so without performing the traditional services.

I have a list of all lands purchased over a 40-year entrepreneurial career as an extreme example of the concentration of peasant land in few hands. The owner inherited a substantial fortune and accumulated a total of 3,300 lots of land totalling twelve to thirteen thousand acres. Over half of this land consisted of sixteen large estates (much of which was uncultivated), and another 1,500 acres were thirty lots over 30 acres each. Most of this latter was planted with coconuts, and all of it was waste land. Most of his land was purchased from private parties rather than from the government. Only 1,621 acres appear to have been bought as crown land. At least 2,500 acres of the larger lots were purchased from private parties in a developed state. I have no record of the development of the remainder of the waste land, which suggests that a greater proportion of his investment went into increasing landholdings than into planting.

The remaining nine-tenths of the lots and one-third of the land were small units of village lands, sometimes undivided fractional shares of paddy fields. Almost all of these were at 40 villages in the sub-district in which the family maintained their "country seat," or just beyond its borders. This is consistent with the findings of anthropologists that village lands were tending to concentrate in fewer hands.

Conclusion

This has been an overview of changes in the land system of nineteenth century Ceylon, stressing the consequences of colonial policy. As in India, "the general results of British legislation were... to establish clear proprietary rights in land, to make land an alienable saleable commodity and thus to bring it within the general property system of a capitalist economy" (Bottomore 1971: 139–140). The consequences of the enforcement of these rights—and the accompanying liabilities—are still not very clear. At present it appears that certain classes of Ceylonese—the village lawyer, the minor civil servant, the entrepreneur, and the salaried professional—had advantages in their awareness of their rights and their willingness to exploit the rigid policies of the government. Further research may give us a better understanding of the ways property norms and the material conditions of the people of Ceylon changed.

REFERENCES

AMEER Ali, A. C. L.
1970 "Peasant Agriculture in Ceylon, 1833–1893." Unpublished M. Phil. Thesis. University of London.

BERTOLACCI, Anthony
1817 *A View of the Commercial, Agricultural and Financial Interests of Ceylon.* London: Black Parbury and Allen.

BOTTOMORE, T. B.
1971 *Sociology. A Guide to Problems and Literature.* 2nd ed. London: George Allen and Unwin.

CODRINGTON, Humphrey William
1937 *Ancient Land Tenure and Revenue in Ceylon.* Colombo: Ceylon Government Press.

DE SILVA, Colvin R.
1962 *Ceylon Under the British Occupation, 1795–1833. Its Political, Administrative and Economic Development.* Vol. II. Colombo: Colombo Apothecaries.

DE SILVA, Kingsley M.
1964 "Studies in British Land Policy in Ceylon. Part I," *Ceylon Journal of Historical and Social Studies* VII (January–June): 28–42.

FARMER, Bertram Hughes
1950 "Agriculture in Ceylon," *Geographical Review* XL (January): 42–66.

MILLS, Lennox Algernon
1964 *Ceylon Under British Rule 1795–1932.* London: Frank Cass and Co.

OBEYESEKERE, Gananath.
1967 *Land Tenure in Village Ceylon. A Sociological and Historical Study.* Cambridge: Cambridge University Press.

PEEBLES, Patrick
1973 "The Transformation of a Colonial Elite. The Mudaliyars of Nineteenth Century Ceylon." Unpublished Ph.D. dissertation University of Chicago.

PIERIS, Paulus Edward
1920 *Ceylon and the Portuguese 1505–1658.* Tellipalai, Ceylon: American Ceylon Mission Press.

SARKAR, N. K.
1957 *The Demography of Ceylon, Colombo*; Ceylon Government Press.

SNODGRASS, Donald R.
1966 *Ceylon: an Export Economy in Transition.* Homewood, Ill.: Richard D. Irwin.

University of Ceylon
1973 *History of Ceylon.* Vol. III. *From the Beginning of the Nineteenth Century to 1948.* Peradeniya, Sri Lanka: University of Ceylon Department of History.

WICKREMESINGHE, K. D. P.
1969 *The Law of Partition in Ceylon.* Colombo: M.D. Gunasena.

Contributions to Asian Studies, Vol. IX

The Impact of Population Growth on the Agricultural Practices and Settlement Patterns of the Anuradhapura Veddahs

JAMES BROW

Swarthmore College, Swarthmore, U.S.A.

IN 1891 THE TOTAL population of Anuradhapura Veddahs was 1170. By 1931 it had climbed to 2045, and by 1970 it had reached 6613.[2] This rate of growth has been somewhat greater than that of Anuradhapura District, Sri Lanka, as a whole. In 1891 the Veddahs made up less than 2% of the population of Nuvarakalaviya, as the present Anuradhapura District was then still commonly called. By 1970 they had come to represent close to 3% of the total population of the District. Declining mortality, increasing fertility and in-migration have all contributed to the population increase, both among the Veddahs and in the larger population. The exceptionally rapid rate of growth since 1931 is largely attributable to the drastic reduction of the death rate achieved by the campaign to eradicate malaria in the late 1940s.

The present paper examines the impact of population growth on two areas of Veddah social life: agricultural organization, and patterns of settlement and mobility. It should be emphasized from the outset, however, that the social and economic changes to be described cannot be attributed exclusively to demographic causes. In the period since 1891 Veddah society has been affected by a variety of exogenous factors, among the most influential of which have been legal changes in the terms of land tenure, administrative policies concerning the supervision of agriculture, improved communications, the restoration of the major irrigation works of the Anuradhapura kingdom, and the increasing commercialization and monetization of the economy. Comprehensive analysis of the effects of these momentous changes, which have not of course been independent either of one another or of the demographic developments, cannot be undertaken here. They are not entirely ignored, but they are held in the background of a study the focus of which is very much more restricted. In the first part of the paper I describe the two modes of Veddah agriculture, and I attempt to show why, despite the rapid growth of their population, the Veddahs have made little attempt to intensify production but have continued to give priority to chena (swidden) cultivation over irrigated rice agriculture. Population growth has, however, been accompanied by an extension of the area under irrigation, and in the later part of the paper I shall describe the population movements associated with that extension.

Veddah agriculture

The Veddahs are traditionally reputed to be a tribal group of hunters and gatherers, descended from the pre-Sinhalese aboriginal inhabitants of Sri Lanka and standing outside the caste system of the Sinhalese Buddhist peasantry. In 1970, however, at least in Anuradhapura District, they were found to fit comfortably into the local caste structure. They speak Sinhalese, practice Buddhism, employ members of the service castes at rites of passage, and subsist by cultivation. As is the customary practice in the Dry Zone of Sri Lanka, the Anuradhapura Veddahs, who form a single caste group (*variga*), inhabit their own separate villages. Their 46 villages are widely dispersed among those of other castes, but with only five exceptions, they all lie within the boundaries of Anuradhapura District. The Veddahs make their living mainly from irrigated rice agriculture and chena cultivation. Chenas are worked by independent households, defined by the presence of separate cooking hearths, on jungle land that is equally accessible to all. Irrigated paddy land, by contrast, is individually held, and its cultivation often requires greater resources than can be mustered within a single household. Paddy cultivation thus provides the economic context in which groups emerge that are larger than the household. The latter is everywhere the unit of consumption. Veddahs work not only in each other's paddy fields, but also in those of neighbouring and more prosperous villages. Labourers were formerly compensated in kind, but today they usually receive a daily wage. Very few Veddahs now work paddy land outside their own villages on share-cropping tenancy (*ande*). Wage labor is a vital source of income for the Veddahs, who are the poorest community in the District, and whose village tanks are generally smaller and less reliable than those of other castes. A small minority of Veddahs keep buffaloes and cattle, the former for use in paddy cultivation, the latter for sale to the Muslim traders who supply the urban beef markets. Goats are also raised in some villages. These subsidiary activities, along with hunting, which nowadays is of very minor significance in the Veddah economy, will be largely ignored here.

We do not know how long the Veddahs in this area have been settled agriculturalists. Members of the present generation are inclined to assert that their grandparents lived mainly from hunting, and there is no doubt that game of all kinds was more plentiful in an age when a single British army officer

---

1 The research upon which this paper is based was carried out between 1968 and 1970 under a grant from the Foreign Currency Program of the Smithsonian Institution (SFCP-45), supplemented by a Dissertation Improvement Grant from the National Science Foundation.

2 Population figures for 1970 are derived from my own research. All earlier figures are from the official Census Reports of the Government of Sri Lanka (formerly Ceylon).

could claim to have shot more than 1200 wild elephants (Tennent 1859: 2: 323; cf. Gooneratne 1970, where Sri Lanka's present elephant population is estimated at 1000). But 19th century reports refer to the Anuradhapura Veddahs as "domesticated Veddahs" (Brodie 1883; Government Agent quoted in Leach 1961: 75), which in the normal usage of the times implies chena if not also paddy cultivation. The relative importance of these occupations a hundred years ago remains uncertain. An Assistant Government Agent in Vavuniya wrote that the Veddahs "are entirely dependent on hunting and occasional chena cultivation" (quoted in Lewis 1895), while according to Parker "a few only cultivate paddy, the rest live by chena cultivation, hunting and honey-collection, especially the latter" (1887: 15). On the other hand, official Grain Tax Registers indicate that in the 1870's at least some Veddahs were growing substantial amounts of paddy (see, for example, Ceylon National Archives file 29, volume 27, which includes figures for the Veddah village of Kukulewa).

Unfortunately, as Leach has demonstrated (1961: 171–2 *et passim*), official records do not permit a precisely quantified assessment of the actual extent of cultivation. It is, however, possible to attempt a qualitative analysis of the relationship between paddy and chena cultivation in the District as a whole, which will indicate how much, or how little, that relationship has been altered by the social, economic and demographic changes of the last eighty years. The general analysis will apply to the specific case of the Veddahs with the single qualification that the inferiority of their village tanks has always obliged them to depend more heavily on their chenas.[3]

Nineteenth century accounts of the economy of Anuradhapura District, and of the North Central Province within which it is contained, are not always in agreement. For example, in his *Manual of the North Central Province*, published in 1899, Ievers states that "the staple food of the mass of the population is *kurakkan*" (1899: 193), *kurakkan*, or finger millet (*Elesuine coracana*) being the main chena product; but according to the Census Report of 1901, only two years later, the staple food in the North Central Province was rice (Census 1901: 70). One source of confusion has been the failure consistently to distinguish between the staple food and the staple product. It is evident that rice was being exported from Anuradhapura District in the nineteenth century (Patterson 1870; Ievers 1887), and this fact seems to have tempted even some recent historians (e.g., Bastiampillai 1970: 10) to infer that local consumption needs were also being amply met. But the more discerning contemporary observers were well aware that the amount of paddy being exported from the District gave little indication of how much was being consumed within it. Ievers notes in his Administrative Report for 1887 that much rice was exported from the District, but he goes on to estimate "that half the population lives entirely on chena-grown grain." The villagers were selling most of their rice to traders from Jaffna, Trincomalee, Matale and Kurunegala, for money and for salt, salt fish, cloth, coconuts and arecanuts.

3 A detailed and quantified analysis of the economic organization of one Veddah village (Kukulewa) will appear in the future.

This trade was long established. Indeed Robert Knox himself engaged in it in the seventeenth century, and actually made his escape through Nuvarakalaviya in the guise of a peddler, carrying "such wares as were vendible in those parts, as Tobacco, Pepper, Garlick, Combs, all sorts of Iron Ware, etc." (Knox 1911: 246). A crucial aspect of this trade was noted by Lieutenant Thomas Nagel, whose *Account of the Vanni* was written in 1793. Nagel reports that "the Jaffna Pedlers who go about with bags... give them (i.e., the villagers) their articles at a year's credit in order to receive in return at the next harvest paddy at the rate of eight stivers per parra" (Nagel 1948: 74; see also Knox 1911: 239–240). In other words, the amount of paddy exported from Anuradhapura District may suggest the extent of rural indebtedness, but cannot be taken to indicate a surplus above local subsistence requirements. This state of affairs has continued into the present century and was noted, for example, in Dyson's Administration Report of 1928 (Dyson 1928).

These reports indicate that credit has been extended more on the security of paddy than of chena crops. Paddy has long been a cash crop, but with the notable exception of gingelly (*Sesamum indicum*, sesame), few chena products have been readily marketable. Freeman noted in 1915 that Anuradhapura District ought to export chena vegetables, "but the villagers, except those near the town of Anuradhapura, find transport prohibitive, and do not know of markets" (Freeman 1915). On the other hand, three years later he noted that some chena vegetables *were* being exported to Colombo and elsewhere by rail (Freeman 1918). But it is clear that, on the whole, the traders and creditors who travelled to the villages came away with paddy rather than chena produce. It is somewhat surprising, therefore, to find that the conditions obtaining in Pul Eliya in 1954 (Leach 1961) were almost exactly the reverse of this. One might have anticipated that the combined effect of the government's Guaranteed Price Scheme, which is intended to stimulate rice production by offering the cultivator a high guaranteed price for his paddy, and the ration program, whereby the consumer obtains some of his rice either free or well below the market price, both of which programs have been instituted since the Second World War, would, if anything, have increased the proportion of paddy produced in the District which reaches the market. But this was apparently not the case in Pul Eliya, at least in 1954. Leach presents the village as a "small peasant community subsisting by the cultivation of rice in irrigated fields of fixed size and position" (1961: 53). He recognizes that in years when there has been a rice crop failure the *kurakkan* grown in chenas may be a very important standby item of diet, but in general the Pul Eliya villager "looks upon this shifting cultivation as his main means of earning a cash income in contrast to his ordinary activities as a rice farmer which provide him with a subsistence living" (1961: 63–64; see also ibid: 289). The historical evidence to which I have referred suggests that if this is the state of affairs in Pul Eliya, then either it is a recent development or the village is exceptionally well favored for paddy cultivation.[4] But clearly one critical question to answer is how often the paddy crop does in fact fail.

Rainfall

Successful paddy cultivation depends upon an adequate local rainfall. The village tanks of Anuradhapura District are not usually connected to any of the major irrigation systems and do not receive water from outside the District nor, indeed, from outside their own local catchment areas. The reliability of a tank is therefore a function of its location since, other things being equal, a well placed tank with a large catchment area will collect more water than a tank with a small catchment area. Pul Eliya evidently is well placed in this respect, for Leach notes that the tank "very seldom dries up completely even in years of drought" (1961: 18). Almost all the Veddah villages, by contrast, occupy the other end of the scale, having unreliable tanks with small catchment areas. These variations are offset by the fact that the larger and more reliable tanks are made to irrigate more extensive tracts of paddy land and to support larger populations. When the rainfall is deficient, however, the fields under a small tank may be impossible to irrigate at all, while the inhabitants of a better endowed village may yet be able to cultivate a reduced proportion of their holdings, a practice which is institutionalised under the name of *bethma* (Leach 1961: 169–171; Pieris 1956: 239–240; Farmer 1957: 45).

Another source of variation is the distribution of the rainfall. When the rains do come, they do not blanket the whole area but are concentrated at a particular time in particular places, so that while an inch or two of rain may fall within an hour in one village, another village a few miles away may receive no rain at all. Over a period of years such variations may be expected to cancel one another out, but it is not certain that they do so during the course of a single season.

Another factor to be considered is the steady improvement in the tanks themselves, which has taken place in the hundred years since the government seriously began to address itself to the restoration and refurbishment of village irrigation works. Clearly the construction of firm bunds and the provision of adequate sluices and spills promote the more efficient use of the available water, and to that extent the irrigation of a constant extent of paddy land should require less water to be stored in the tank. In fact, as will be discussed, the improvement of the irrigation works has been accompanied by an extension of the area under cultivation and an increase in the local population to be supported, so that the amount of rainfall required has in no sense declined.

Bearing these considerations in mind, it is useful now to examine the rainfall statistics for the area in some detail.[5] Table I reports the rainfall recorded at Anuradhapura for every month between January 1871 and December 1897. I have selected this period because the Government Agents of the time were

4 Leach specifically denies any claim that Pul Eliya is a typical Nuvarakalaviya village (1961:14).

5 For a geographer's analysis of rainfall and water-supply in the Dry Zone of Sri Lanka, and particularly for a useful assessment of the Zone's "effective rainfall", see Farmer (1956 and 1957:22–31).

usually concerned to report the overall success or failure of each season's cultivation (see Administration Reports, 1871–1897). It will be seen that there was a great deal of variation in the annual rainfall recorded, ranging from 79.86 inches in 1885 to 32.58 inches in 1873, with an average of 55.16 inches.

Table I

Monthly Rainfall in inches at Anuradhapura, 1871–1897

Year	*Jan.*	*Feb.*	*Mar.*	*Apr.*	*May*	*Jun.*	*Jul.*	*Aug.*	*Sep.*	*Oct.*	*Nov.*	*Dec.*	*Total*
1871	11.29	3.71	2.87	13.24	2.14	—	—	1.82	1.82	4.81	19.15	7.29	68.14
1872	1.49	0.27	2.87	9.89	3.05	0.03	0.72	0.11	7.10	3.67	17.41	2.96	49.57
1873	0.20	2.90	1.11	4.51	3.42	0.40	0.03	4.49	—	4.53	6.12	4.87	32.58
1874	1.29	3.33	1.21	5.41	3.06	0.12	—	6.24	7.84	2.65	7.60	7.17	45.92
1875	0.69	0.02	3.42	8.45	3.96	5.27	0.03	2.01	2.82	10.60	11.27	8.27	56.81
1876	1.86	0.37	3.09	5.42	1.81	0.52	0.03	2.75	4.42	4.49	5.28	4.62	34.74
1877	0.25	—	4.61	9.01	4.52	1.03	0.05	0.44	2.89	10.69	18.35	19.97	71.81
1878	7.46	—	1.86	10.32	4.34	0.38	3.21	0.35	2.70	5.24	11.25	3.98	51.09
1879	1.72	1.10	7.41	3.73	4.53	—	7.14	3.53	2.51	8.88	15.07	7.64	63.26
1880	3.85	1.14	6.91	7.11	0.03	0.04	—	1.22	0.02	9.54	16.81	4.03	50.70
1881	5.35	0.03	1.57	6.62	2.99	0.22	0.40	0.44	4.18	3.37	12.21	12.27	49.65
1882	5.23	2.28	3.05	3.17	2.93	1.73	0.42	5.87	0.02	8.24	8.88	20.36	62.23
1883	1.94	5.58	3.01	6.81	7.02	8.53	0.72	4.54	—	12.02	16.47	8.28	74.92
1884	0.93	—	1.59	5.64	3.94	1.28	—	0.11	4.10	15.58	15.21	8.36	56.74
1885	0.74	0.37	4.73	9.45	8.79	7.29	—	0.48	4.45	15.69	16.24	11.63	79.86
1886	0.58	1.08	1.24	6.85	4.46	3.04	3.66	5.87	6.00	6.88	6.33	2.52	48.51
1887	0.86	2.72	2.21	11.45	—	0.91	—	0.29	—	15.58	6.46	26.53	67.01
1888	0.10	—	—	4.10	8.55	1.04	—	—	5.59	8.63	5.16	7.88	41.05
1889	0.86	—	1.54	10.73	3.72	0.06	4.62	1.23	5.96	8.58	9.16	3.30	49.76
1890	2.82	2.21	1.21	18.02	0.38	1.28	0.11	0.79	0.15	3.14	5.31	5.37	40.77
1891	0.50	1.14	6.41	6.44	20.33	0.53	0.28	—	0.25	19.98	7.85	10.93	74.64
1892	6.53	4.73	0.57	8.73	0.03	0.72	0.67	2.78	0.55	2.64	11.94	8.52	48.41
1893	0.72	0.64	6.41	4.71	0.49	0.58	4.40	0.03	0.08	8.79	8.12	4.88	39.85
1894	2.83	0.39	3.01	6.10	0.26	0.08	0.68	1.42	1.44	11.75	11.56	6.58	46.12
1895	3.21	—	0.99	11.38	0.06	2.33	0.04	4.39	1.16	18.90	7.83	9.50	59.79
1896	4.25	0.39	4.26	4.39	3.97	6.46	1.48	0.12	2.85	14.53	14.02	12.82	69.54
1897	6.92	1.21	0.67	7.13	1.15	3.44	0.10	1.83	6.67	4.08	8.29	18.28	59.77
Average	2.76	1.39	2.88	7.73	3.70	1.53	1.07	1.97	2.80	9.02	11.09	9.22	55.16

More than half the mean annual rainfall comes in October, November, and December, and it is the rain collected at this time, which is roughly the period of the North East monsoon, that is used to irrigate the main (*Maha*) crop. More than half of the remainder of the rain normally falls between March and May, and this sometimes permits a second cultivation season (*Yala*). *Yala* rains, however, are rarely sufficient in themselves, and cultivation is often possible only if the previous *Maha* rainfall has been abundant, and a surplus of water remains in the tank. An alternative scheme is to cultivate only a fraction of the fields in the *Maha* season, and to conserve water for a later, *Yala* crop (Leach 1961: 52–53). The unreliable Veddah tanks only rarely permit *Yala* cultivation.

Yala cultivation is, then, generally subordinate to *Maha* cultivation, and the most critical period, when rain must fall if cultivation is to be successful, is that between October and January. Water cannot be stored indefinitely in the open tanks because of what Farmer describes as an "astonishingly high" rate of evaporation in the dry season; he reports a figure of half an inch daily (1957: 25–26). Government Agents reported on the success or failure of *Maha* cultivation in eighteen of the twenty-six seasons between 1871 and 1897. In all ten of the seasons which they declared successful the recorded rainfall between October and January (inclusive) was more than thirty-two inches. In all eight of the unsuccessful seasons the rainfall during the same period was less than twenty-five inches (Table II). It seems reasonable to infer, therefore, as a crude index, that successful cultivation requires a minimum of between twenty-five and thirty-two inches of rainfall during these four months. Fortunately for this inference the rainfall recorded between October and January in the years when the Government Agent failed to report on cultivation was always either more than thirty-two or less than twenty-five inches. Applying

Table II

Rainfall recorded at Anuradhapura between October and January each season between 1871–72 and 1896–97 and the success or failure of paddy cultivation in that season

Season	*Rainfall Oct–Jan (Inclusive)*	*Success* (S) *or Failure* (F)	*Reported* (R) *or Inferred* (I)
1871–72	32.74	S	R
1872–73	24.24	F	I
1873–74	16.81	F	R
1874–75	18.11	F	I
1875–76	32.00	S	I
1876–77	14.64	F	R
1877–78	56.47	S	I
1878–79	22.19	F	R
1879–80	35.44	S	I
1880–81	33.73	S	R
1881–82	33.08	S	R
1882–83	39.42	S	I
1883–84	37.70	S	I
1884–85	39.89	S	R
1885–86	44.14	S	R
1886–87	16.59	F	R
1887–88	48.67	S	R
1888–89	22.53	F	I
1889–90	23.86	F	R
1890–91	14.32	F	R
1891–92	45.28	S	R
1892–93	23.82	F	R
1893–94	24.62	F	R
1894–95	33.10	S	R
1895–96	40.48	S	R
1896–97	48.29	S	R

the crude rainfall index to these eight seasons, the result, for the period as a whole, is a total of fifteen successful years and eleven years of failure. A successful crop, in other words, can be expected in no more than three *Maha* seasons out of five.

This is a conservative estimate, being based only on the frequency of deficient rainfall. But the rains can also be excessive. Heavy rainfall during a concentrated period can cause floods which breach bunds and destroy crops. Between 1871 and 1897 this is only recorded as having occurred on a large scale in December 1882, but it is likely that at least some tanks were breached in other comparable months, such as December 1877 and December 1887. Moreover I have taken no account of other hazards in the natural environment, such as crop blights or the devastating incursions of elephants and other pests.

It seems improbable, in the light of this estimate, that more than a very few exceptionally fortunate communities, of which Pul Eliya may be one, have been regularly able to derive their subsistence from irrigated paddy cultivation. Even if yields in the successful years were startingly high, the surpluses thus generated would hardly have been sufficient to see the villagers through an unpredictable succession of bad years. Five of the six seasons between 1888 and 1894, for example, were unsuccessful. The Veddahs in particular, with their tanks of less than average reliability, cannot have depended on wet-rice cultivation for their subsistence.

The Anuradhapura District villager does not, of course, need to be told that paddy cultivation is an extremely precarious business, but observers and officials who, if not foreigners, have usually come from other parts of the island, seem to have been reluctant to learn this lesson. The thrust of agrarian policy has nearly always been to discourage chena cultivation in favour of more intensive paddy cultivation. Now chena crops can also suffer from a drought, but they suffer much less from low rainfall than does a paddy crop. The peasant accordingly gives priority to his chena, which he prudently sees as the least uncertain means to procure his subsistence. Officials, on the other hand, have consistently emphasized the disadvantages of chena cultivation, and the literature abounds with more or less prejudiced elaborations of the criticisms already made by Rhys Davids in 1871: "chena cultivation prevents civilized habits and enterprise: it is unwholesome and unhealthy; it is incompatible with paddy or other more renumerative cultivation because the working times clash; it destroys forest resources; and the rotation of the soil prevents any permanent improvement of the land" (Rhys Davids 1871: 93). Rhys Davids did also recognize, more sympathetically than some who followed him, that the villagers had good economic grounds for preferring chena cultivation, and were not simply motivated by their enjoyment of "the destruction of the forest (and) the fine blaze after it" (ibid: 91), or by their conservative addiction to irrational custom, but this recognition has prompted only reluctant gestures intended to soften the impact of a policy which would severely restrict chena cultivation, or even eliminate it altogether. As Leach says, "shifting cultivation... is a matter on which villagers and the administration hold diametrically contrasted points of view" (1961 : 289; see also ibid: 61–64). In practice, official dogma has frequently been forced to yield to economic necessity, and chena cultivation, despite administrative harassment, has retained the villager's loyalty.

Accordingly, attempts in recent years to intensify paddy cultivation, as distinct from attempts to extend it, have made slow progress. The government's efforts to increase rice production by encouraging transplanting, the use of fertilizers and insecticides, systematic weeding, etc., are seen by the villager as an improvident invitation to gamble more of his resources than he can afford on a most uncertain outcome. It does not seem likely that the greater expenses of intensive cultivation will be willingly incurred by the villager until the government can either guarantee an adequate water-supply (presumably by bringing in water from the Wet Zone) or provide an attractive and workable scheme of crop insurance, or alternatively, as an even less realistic condition, until the cultivator has accumulated sufficient reserves to be able to risk several successive failures before he recoups his losses.

Rhys Davids' assessment (1871: 85), that arable land in Nuvarakalaviya is cultivated on average in two years out of five, not only gives some support to my own estimate but also draws attention to another expression of the cultivator's prudence. Unsuccessful seasons are often marked not so much by crop failure as by the failure to cultivate at all. Here again the government and the cultivator have frequently clashed, and for the same reasons that I have just adduced. Farmer supplies an authoritative defence of peasant practice in this respect:

> "The peasant is... often criticized for waiting until his tank is full before starting *Maha* (north-east monsoon) cultivation, thus wasting the direct effect of early rains which occur often enough to permit preliminary cultivation without irrigation. But this practice is, in reality, like so many apparently ill based indigenous customs, a sensible reaction to local circumstances; for if, encouraged by early rains, the peasant cultivates before his tank is full, he stands to lose both effort and seed if later rains fail" (1957:45).

The villager is in any case busy at this time with the chena on which his subsistence more fundamentally depends.

Chena and Paddy

Chenas are cultivated in the scrub jungle that separates the villages from one another. In most parts of the District villages are between half a mile and two miles apart. The chenas themselves are usually between one and three acres in extent. They are normally cultivated for two years and then abandoned. It is difficult to determine the average length of time which is then allowed for the jungle to regenerate, and thus the extent of erosion. Farmer notes that the peasant "is aware of the insufficiency of the short period of regeneration (e.g., seven years) enforced by a growing population" (1957: 48), and Veddahs concur that this period is now both common and insufficient. But Farmer also refers to the work of Joachim and Kandiah (1948) which shows "that under the system practised in these regions chena cultivation depends almost entirely on nitrogen and nutrients made available by the burning of the jungle, soil reserves remaining largely untapped and in any case being readily restored after five to ten years of regeneration" (Farmer 1957: 48; c.f. Conklin 1961). Farmer considers that "under the physiographic and demographic conditions which have prevailed since (ancient times) in most parts of Nuwarakalaviya... it (i.e., chena cultivation) has enabled high land to be used for very many centuries without impairing the soil" (*idem*). The maintenance of this regime has up to now been threatened less by population growth than by the official policy that was intended to protect it. Leach comments on the government regulation, designed to conserve forest resources, which requires that:

> "where land is cleared for shifting cultivation it must be jungle which is 'not more than ten years old'! This implies that the fertility of this cleared ground shall be as low as possible and it also, incidentally, makes it certain that the resulting erosion shall be the maximum possible" (1961:62).

With regard to population, Farmer cautiously estimates that the critical density in Nuwarakalaviya, where the villager also undertakes paddy cultivation, is about 200–250 per square mile (1957: 49). Even if this estimate proves optimistic it is still four times the population density recorded in 1953 (Abhayaratne and Jayewardene 1967: 45), and at least twice that of today.

Of course the population is not evenly distributed within the District, and chena land is therefore not equally available everywhere. In this respect some Veddah villages have felt the impact of increased population density more than others. In particular the villages of Dambagahaulpota and Migahapattiya, which have been displaced by the restoration of the ancient tank at Huruluwewa, and Rotawewa, which is immediately adjacent to two other colonization schemes, have suffered a serious reduction of their chena lands. On the other hand the small Veddah villages in Willachchiya, north-west of Anuradhapura town, which remains an area of sparse population, are still surrounded by relatively extensive tracts of jungle available for chena cultivation.

But while the prospects for chena cultivation vary somewhat among the Veddah villages, within any one village all inhabitants enjoy equal access to the surrounding jungle. As I have already noted, chenas are grown on Crown land which, with very few exceptions, has not become alienated to individuals. It is recognized that a man who has cleared a particular plot has the right to work it again in the following year, but the individual possession of jungle plots does not extend beyond this. At various times the government has attempted to restrict chena cultivation to the poorest class of villagers, but to the limited extent that this policy has ever been effective, it has discriminated little among the Veddahs, virtually all of whom are, even by local standards, extremely poor. It is possible that regulations restricting the permissible size of chenas have been more successful, but if so, this is probably only partly due to the regulations themselves. Rather, it has been the steadily increasing availability of wage labour outside their own villages, and, since at least as early as the 1870's, the provision of relief in years of acute distress that have somewhat diversified the Veddah's defences against famine. If the Veddahs formerly grew larger chenas than they needed in normal years in order that sufficient crops would survive in years of drought, it may be that they have now come to rely on wage labour and relief measures to see them through the bad years. But this is speculation. It may equally be that they grew smaller chenas in an age when game was more plentiful.

In summary, chena cultivation is egalitarian. It is undertaken by the smallest continuously cooperating economic unit, the household, on land that is not individually owned and which is equally accessible to all. The capital requirements are minimal, an axe, a mamoty (hoe) and a sickle being the only necessary tools. Official regulations aside, it is restricted only by the amount of manpower that the individual household can muster. It is largely geared to subsistence and provides few opportunities either for monetary profit or for the elevation of one villager above his fellows. In all these respects, as we shall now see, it differs from paddy cultivation.

Although the Veddahs do not practice intensive wet rice cultivation, their paddy crop still requires a greater input, quite apart from the investment in the irrigation works, than does chena cultivation. The return from seed is about half that of *kurakkan*, and seed is more expensive. The necessary agricultural tools are also more expensive, as well as more numerous. If buffaloes are not owned they must be hired for certain operations and it is, moreover, commonly the case that at certain times a greater concentration of labour is required than can be mobilized within a single household. Cultivators in different households therefore enter into a variety of relationships with one another whereby labour is exchanged either for payment in kind, or for wages, or for an equivalent return of labour at some other time. In addition the private tenure of paddy land gives rise to a complex system of relations among those who have an interest in a particular plot of land, between owner and tenant, mortgager and mortgagee, employer and employee. In contrast to the independent household organization of chena cultivation, paddy thus supports a dense network of social relations among members of different households. These relations reinforce the conditions fundamentally laid down by common interest in the village irrigation system, which requires close cooperation among the cultivators while providing ample opportunity for conflict.

But the fact that paddy cultivation is expensive and its outcome uncertain does not make it unprofitable. On the contrary, a successful harvest brings great rewards, while the risk of failure can be minimized. I have already mentioned the main precaution that the villagers take, that of waiting until the tank is full before beginning to cultivate. This does not ensure success, but it does limit losses.

Paddy land is therefore much sought after. But at any one time the amount of paddy land in a village is close to the limits imposed by the state of the irrigation works and the available water supply. If a man is to increase his holdings, then, he can often only do so at the expense of others. This happens frequently enough, for although inheritance is the only normatively approved mode of transfer, sales, mortgages and tenancies are common. Such transfers are of crucial importance in the internal processes of village life, but they fall outside the scope of the present account, which takes the Veddah community as its unit of analysis, rather than the individual members of that community.

The other way in which paddy land can be acquired is by bringing previously dry land under irrigation, a process that is known in Sinhalese English as "asweddumizing." Prior to 1935 the development of new paddy lands was largely left to the initiative of individual villagers, but in recent years it has been achieved with administrative encouragement and assistance. The main thrust of official policies has been to restore and refurbish the vast irrigation works of the ancient Anuradhapura civilization, and to settle colonists under them (Farmer 1957). With the few exceptions mentioned earlier, these projects have had little direct impact on the productive organization of the Veddahs, for the village tanks are independent of the major irrigation schemes and very few Veddahs have been granted land in any of the colonies. But the government has also attempted to improve the village systems in recent years, and has thereby contributed to a rapid increase in the area under irrigation.

An idea of the magnitude of this change can be obtained by comparing the major categories of land in the villages. *Paraveni* property includes house-sites, gardens, and paddy fields which were under cultivation when the colonial government surveyed the area at the beginning of the present century. Such lands were treated as ancestral private property. All other land, and most village tanks, were deemed to be Crown property. Subsequently villagers were enabled to purchase plots of Crown land either for new house-sites or in order to asweddumize them. Land thus acquired is freehold, and is called *sinakkera*. Since the Land Development Ordinance of 1935, however, it has no longer been possible to acquire *sinakkera* land from the government. Crown land is now granted to villagers only on permanent lease (*badu*).

An accurate assessment of the amount of land held under each of these forms of tenure requires a more exhaustive study of official records than I have been able to undertake for all the Veddah villages. But an initial survey suggests that the experience of Kukulewa, the village whose land history has been studied in some detail, is not unrepresentative. Including only plots that were cultivated at least once between 1968 and 1970[6] the present extent of paddy land in Kukulewa comprises 33 acres of *paraveni* land, 33 acres of *sinakkera*, and 64 acres of *badu* land. This expansion of cultivable land can be compared with the growth of population in Kukulewa from 108 in 1891 to 207 in 1931, to 552 in 1970. This reveals a lowering of the *per capita* holding of paddy land

6 This restriction excludes land that has been acquired with the intention of bringing it under irrigation but which has not yet been asweddumized.

from roughly 0.3 acres in the period before 1931 to 0.24 acres in 1970. My estimate for the totality of Veddah villages is that the average *per capita* holding was about 0.3 acres in both 1931 and 1971. An estimate of holdings in 1891 is harder to derive without more extensive documentary research, owing to the villagers' tendency to represent their *sinakkera* lands as *paraveni*.

For the district as a whole, it was officially estimated at the end of the nineteenth century that between 40,000 and 50,000 acres were under paddy cultivation (Fisher 1885, Ievers 1899). In 1952 the estimate was that 115,670 acres were asweddumized, of which 78,860 were cultivated in the previous season (Rajendra 1952). Between 1901 and 1953 the population of the District rose from 79,100 to 229, 282.

These figures should be accepted only with caution, but there is no reason to suppose that the general picture they present is grossly inaccurate. The Veddah community, like the District as a whole, has experienced a rapid growth in its population over the last eighty years. But the extension of the area under irrigation has more or less kept pace with the population growth, and per capita holdings of paddy land are scarcely less now than they were in the past. The expansion of cultivation has, however, involved a considerable redistribution of the population, and it is to this process that I now turn.

Population Distribution

New land has been brought under irrigation both in pre-existing Veddah villages and in newly established ones. In 1891 the Veddahs inhabited 27 villages. In 1970 there were 46 Veddah villages (Table III). The combined population of the villages established since 1891 is 1973, which is 29.7% of the present total, and these villages contain approximately one third of the irrigated land in Veddah hands.

Table III
Size and Distribution of the Veddah Population

	1891	*1931*	*1970*
Number of villages	27	35	46
Total population	1170	2045	6613
Average village population	43.3	58.4	143.8

Although there are now 19 more Veddah villages than there were in 1891, the Veddahs have actually established 31 new villages during the period; that is to say, they have also abandoned 12 villages since 1891. Veddah informants usually explain that these villages were abandoned because of the persistence of infectious disease within them, but there is evidence that in most cases this is a retrospective rationalization that serves to obscure the more decisive fact that abandonment has followed the loss of control over village lands to members of other *variga*. Veddah expansion is therefore to be seen as occurring within the more general process of their displacement. Displacement is not directly related to population growth within the Veddah community, although it may well be that those other caste or ethnic groups that have obtained possession of tanks and villages that were formerly in Veddah hands were moved to do so by the pressure of their own increasing numbers on the resources available to them.

The Veddahs themselves have not acquired possession of any villages in this way. Their new settlements have all been pioneering attempts to establish new communities in the jungle. They have, however, been facilitated by the existence within the District of ancient and long abandoned village tanks dating from the time of the Anuradhapura kings. Modest refurbishment of these irrigation works has been sufficient to bring some land under cultivation.

Five of the new Veddah villages have been established by displaced communities that have moved *en masse* to pioneer new settlements. In six of the other seven cases of displacement, the population of the abandoned village was dispersed among nearby Veddah villages. In the twelfth case, the displaced community divided into three groups, all of which independently pioneered new settlements. Each of these three groups was headed by a middle-aged man and was composed of himself, his wife, children, children's spouses, and grandchildren.

This is also the common form of the pioneering group in instances where new settlements have been founded by segmentation of an ongoing village community. In all but one of the 23 cases of segmentation, it has been a middle-aged man, accompanied by his wife and children, at least one of whom has been old enough to share the initial heavy work, who has split off from his former village to pioneer the new settlement.

Most new villages are established within three or four miles of the parent community. Their founders do not immediately break all social ties with their kinsmen and erstwhile neighbours, and many continue to work paddy land that they have retained in their native villages. But it is difficult for a man to live in one village and to work land in another that is more than a couple of miles away. And most holdings are too small for a non-resident proprietor to find it worthwhile to supervise cultivation by a tenant and to ensure that the obligations of *rajakariya* (literally, "king's duty", i.e., irrigation channel maintenance, tank bund repair, fencing, guarding the fields at night, etc., all of which are communal responsibilities that remain with the proprietor) are properly performed. If a man's holdings are large enough to make this worthwhile, he is most unlikely to pioneer a new village.

The opportunity to found new villages has been differentially available within Anuradhapura District, especially in the period since 1931. It is only in the area known as Willachchiya, to the northwest of Anuradhapura town, where tanks and villages are smaller and the overall population less dense than elsewhere, that this option has remained open. Seventeen of the 31 Veddah villages founded since 1891 have been established since 1931 in Willachchiya. Thus in Willachchiya much of the population explosion has been absorbed in newly founded villages. In 1891 there were only five Veddah villages in Willachchiya, and their combined population was 94 (average 18.8 per village). By 1931 there were nine villages with a combined population of 224 (average 25.3), and by 1970 there were twenty-one villages with a combined population of 813 (average 38.7). Thus although the present population is nearly nine times greater than it was eighty years ago, the population of the average village has barely doubled. More than half the present Veddah population in Willachchiya and more than half of the presently cultivated paddy land are located in villages that have been founded during the last forty years.[7]

Elsewhere in the District the opportunity to establish new villages has been inhibited by the greater press of the general population, and only one new village has been founded since 1931. Thus, although the rate of population growth has been less than in Willachchiya, the average village population has risen much more rapidly. In 1891 there were twenty-two Veddah villages outside Willachchiya, with a combined population of 1076 (average 48.9 inhabitants per village). By 1931 there were twenty-six villages and the population had grown to 1821 (average 70.0). In 1970 there were twenty-five villages and a population of 5800 (average 232.0).

The faster rate of population growth in Willachchiya is attributable not to a higher rate of natural increase but to migration differences. Residential mobility among the Veddahs is largely confined to marriage and the establishment of new villages. Excluding those who have pioneered new settlements, more than 96% of married Veddahs were living either virilocally or uxorilocally in 1970. The rule of marital residence is ambilocal, and the overwhelming majority of residential moves occur when a man or woman goes to live in his or her spouse's village. Many couples have taken advantage of the residential option to settle in Willachchiya, where the opportunities for acquiring new paddy land and for unrestricted chena cultivation have been greatest. Between 1931 and 1970 men and women raised in the Veddah villages of Willachchiya contracted 101 marriages with people from outside Willachchiya, of which 76 were followed by residence within the Willachchiya Veddah community. In other words, for every man or woman who has left the villages at marriage, three others have come in.

Thus the residence rule functions to redistribute personnel among the available resources. Residential choices can also be explained causally, in terms of a couple's common, but not always overriding, decision to settle where its prospects are greater.

---

7 My category of "Willachchiya Vedda Villages" includes all the Veddah villages located within the former Willachchiya and Galkadawala tulanas, now Gramasevaka's divisions numbers one and fifteen in Anuradhapura district. I have also included one village (Magurahitiyawa) which is actually half a mile over the border in Pahala Wew tulana, but which, on account of its proximity, intimate connections and similar character, deserves to be grouped with the other Willachchiya villages.

Conclusion

The possibility of bringing new land under irrigation and the continued availability of jungle land for chena cultivation have enabled the Anuradhapura Veddahs to respond to rapid population growth without significantly altering their preferred allocation of resources between the two modes of agriculture, and without adopting the more intensive techniques of wet-rice agriculture that outside authorities have urged on them. It is doubtful, however, that they will be able to maintain this strategy much longer. As their population continues to grow, the chena cycle will become dangerously short, while even in Willachchiya it appears that favourable locations for bringing more land under irrigation are becoming scarce.

In this paper I have tried to show why the Veddahs allocate their resources as they do, and I have described certain changes in the distribution of the population that have been associated with the expansion of the area under irrigation. These changes may be seen as responses to population growth, but population growth alone does not adequately explain the particular redistributive movements that have occurred. Thus, as I have shown elsewhere (Brow 1974), the decision to establish a new village is not simply an automatic response to population pressure, but is also an act of secession which expresses a political cleavage in the community and which may mark an important stage in the process of village development. Analysis at the level attempted here, which is that of the village and the community of villages, i.e. the *variga* as a whole, is complementary to a more detailed study of social process at the level of the individuals and groups, especially households, whose purposeful and decisive actions produce the social formations that are observable on the ground. Analysis at this latter level is necessary, for example, to take account of the fact that while the overall balance between chena and irrigated agriculture appears to have changed very little, at any one time there is great variation in the degree to which individual households emphasize one or the other of these two activities. I would suggest that this variation is not simply a function of the unequal distribution of individually held paddy land, but is also conditioned by demographic factors that have not been considered here, such as the age and sex composition of the household. A study of the household economy may also show that the aggregate growth of Veddah population, which is now pressing against the limits of agricultural expansion, may be the result of fertility behaviour that is economically "rational" for the individuals concerned; that is, under present social and economic conditions, it may be advantageous for Veddah couples to produce large families. But the detailed analysis that would test these hypotheses remains to be done.

REFERENCES

BASTIAMPILLAI, B.
1970 "The Revival of Irrigation Enterprise in Ceylon, 1870–1890", *Ceylon Journal of Historical and Social Studies* 10:1–2:1–26.

Brodie, A. O.
1883 "Topographical and Statistical Account of Nuvarakalaviya", *Journal of the Royal Asiatic Society (Ceylon Branch)* 3:9:136–61.
Brow, J. B.
1974 "Veddah Villages: A Study of the History and Social Structure of Veddah Communities in Anuradhapura District, Ceylon." Unpublished PhD. dissertation, University of Washington.
Conklin, H. C.
1961 "The Study of Shifting Cultivation", *Current Anthropology* 2:27–61.
Dyson, E. T.
1928 "Report on the North Central Province", *Administration Reports,* Ceylon Government.
Farmer, B. H.
1956 "Rainfall and Water Supply in the Dry Zone of Ceylon", in C.A. Fisher and R.W. Steel, eds., *Geographical Essays on British Tropical Lands.* London: Phillips.
1957 *Pioneer Peasant Colonization in Ceylon.* London: Oxford University Press.
Fisher, F. C.
1885 "Report on the North Central Province", *Administration Reports,* Ceylon Government.
Freeman, H. R.
1915 "Report on the North Central Province", *Administration Reports,* Ceylon Government.
1918 "Report on the North Central Province", *Administration Reports,* Ceylon Government.
Gooneratne, B. M. W.
1970 "The Ceylon Elephant, Elephas Maximus Zeylanicus: Its Decimation and Fight for Survival", *Ceylon Journal of Historical and Social Studies* 10:1–2:149–160.
Ievers, R. W.
1887 "Report on the North Central Province", *Administration Reports,* Ceylon Government.
1889 *Manual of the North Central Province, Ceylon.* Colombo: G.J.A. Skeen.
Joachim, A. W. R. and S. Kandiah
1948 "The Effect of Shifting (Chena) Cultivation and Subsequent Regeneration of Soil Composition and Structure", *Tropical Agriculturist* 104:3–11.
Knox, R.
1911 *An Historical Relation of the Island Ceylon in the East Indies* (originally 1681). Glasgow: James MacLehase and Sons.
Leach, E. R.
1961 *Pul Eliya: A Village in Ceylon.* Cambridge: Cambridge University Press.
Lewis, J. P.
1895 *Manual of the Wanni Districts.* Colombo: H. C. Cottle.
Nagel, T.
1948 "Account of the Vanni, 1793", *Journal of the Royal Asiatic Society (Ceylon Branch)* 38:106:69–74.
Parker, H.
1887 "The Wanniyas", *The Taprobanian* 2:15–21.
Peterson, G. W.
1870 "Report on Nuvarakalaviya", *Administration Reports,* Ceylon Government.
Pieris, R.
1956 *Sinhalese Social Organization: The Kandyan Period.* Colombo: Ceylon University Press Board.
Rajendra, M.
1952 "Report on the North Central Province", *Administration Reports,* Ceylon Government.
Rhys-Davids, T. W.
1871 "Report on Nuvarakalaviya", *Administration Reports,* Ceylon Government.
Tennent, Sir J. E.
1859 Ceylon: *An Account of the Island, Physical, Historical and Topographical.* Two volumes. London: Longman, Green, Longman, and Roberts.

Contributions to Asian Studies, Vol. IX

Kinship and Land Rights in the Context of Demographic Change[1]

A. J. SELVADURAI

San Diego State University, San Diego, U.S.A.

THIS PAPER examines changes in kinship and land rights in Mulgama, a Sinhalese village in Sri Lanka's Western Province. The main concerns of the paper are twofold. First, it relates these changes to demographic trends within and outside Mulgama; the former are viewed here as re-adjustments within a collective ideology to fit the pressures of population on land, specifically, land used for paddy cultivation. Second, the paper examines how these changes have come about by focusing on ideological, marital, residential, and inheritance strategies of individuals in Mulgama, notably, its "elite", who have acted as the agents of ideological change.

In emphasizing individual strategies as important ingredients of changes in the ideological structure, the paper provides a further perspective on demographic studies on Sri Lanka. Specifically, I draw attention to an observation, which is frequently made in these studies, that population pressures have directly contributed to the process of social disintegration of the village community (see Arulpragasam 1961, Perera 1968, Sarkar 1957, and Sarkar and Tambiah 1957).

The notion of the socially disruptive effects of population increase rests on the assumption that this increase is a sole determinant of change. I take the view that population pressures form the backdrop for change, and that change itself, following Leach (1967), is the total impact of individual actions on a collective ideology. Briefly, Leach takes the position that "Every individual of a society, each in his own interest, endeavours to exploit the situation as he perceives it, and in so doing alters the structure of society itself" (1967: 8). In his view, structural change occurs when there are alternative courses of action and inconsistencies in values which certain individuals may manipulate for their own advantage.

Invariably, those capable of initiating and sustaining significant changes turn out to be the ones having the power (economic, political or other) to back up their innovative strategies. We may regard these individuals as *primary*

---

1 I conducted the research on which this paper is based during the year, 1971–72, when I was attached to the academic staff of the University of Ceylon, Peradeniya; data were gathered for a doctoral dissertation. I am grateful to several colleagues for their criticisms of an earlier version of the paper; in particular, I thank James Brow, Barbara Pillsbury, and Phillip Staniford. The responsibility for opinions expressed here is entirely mine.

agents, and, those who emulate their actions as the *secondary agents* of change.

In Mulgama, the primary change agents were its elites, a handful of politically and economically strong individuals, who used their positions to bring about changes in the nature of land rights. The cumulative result of their actions was a shift in the ideology of kinship and property rights from egalitarian to hierarchic values. Although the "non-elites" (or, villagers without political privilege) have adopted elitist strategies to a considerable extent, most of them have not abandoned "traditional" values, either. The break from tradition is certainly not a sharp and irrevocable one, at least for the non-elites.

The description and analysis will be in two parts: Part I discusses the historical background of population, land, and kinship, and evaluates the impact of population increase on the resources of individual households; Part II describes the strategies of both elites and non-elites in the context of diminished resources, and examines the role of elites as ideological change agents. The paper concludes with a discussion of the larger implications of this ideological change for village society with attention being drawn to such forms of conflict as land disputes and litigation, as well as to the emerging migratory trend from the village to urban areas.

PART I. LAND, POPULATION AND KINSHIP STRUCTURE

Setting

Mulgama is situated on the northern boundary of the Western Province The village is a part of Udugaha Pattuwa, a revenue division of Kalutara District. In 1971, Mulgama's population was 1963 persons forming 350 different households. The people are members of the "farmer" (*goigama*) caste, and are Buddhists.

Mulgama has many features which may be considered "typical" of villages in the general area. A large majority of its people are subsistence paddy cultivators, although, in 1971, there were about 5 households which had enough paddy to sell after keeping the amounts required for seed and consumption. In addition to rice-farming, many are engaged in other occupations, which, unlike subsistence cultivation, do bring them cash incomes. Almost every household has a few rubber trees in its compound; they require less or very little attention when compared to a paddy crop, and each tree gives about 10 rupees worth of latex a month. Villagers also grow coconut, assorted vegetables and yams; they sell any surplus from these crops at the fairs held once a week in the neighboring towns. Mulgama also has 7 schoolteachers, 3 government clerks, one policeman, 4 carpenters, 6 masons, 4 traders, and about 20 non-agricultural laborers; all these persons work outside the village, a few in the city of Colombo which is about 21 miles away.

The dependence on the world outside for cash income, which the villager needs for purchasing consumer goods, for educating his children and, sometimes, for litigation, has in no way diminished his outlook on paddy cultivation as the one, stable basis of subsistence; besides, resources outside the village are governed by factors beyond his control. The importance of paddy is also reflected in the fact that those getting cash incomes have invested some of their earnings in new machinery, and in such inputs as chemical fertilizer, weedicides and herbicides for increasing yields. However, a majority do not have the capital to adopt these agricultural innovations.

Mulgama farmers have an adequate water-supply to cultivate two seasons: *yala* (April through August) and *maha* (September through January). The Wet Zone as a whole receives an annual average rainfall of over eighty inches; Mulgama fields are also irrigated by a steady stream originating about 12 miles north of the village. As I will indicate later on, the scarce resource is not water, but paddyland itself.

The unit of land ownership is a share. A person generally inherits a share of the parental property, or, he may acquire other shares through cash purchase. Shares are also given out on lease or on a sharecropping basis. A person has a share or shares in a named plot or plots of land. In Mulgama, there are about thirty-one different plots, each with its formal boundaries[2]; they form a continuous stretch of paddyland.

Land and Population

A noteworthy feature of the setting is the obvious imbalance between the lowland area under paddy cultivation and the area in highland. Out of a total of about 650 acres, which make up Mulgama's total land area, a mere 25% (or 163 acres) is used for paddy, and the remainder for residence and for the cultivation of highland crops. This land use pattern may be attributed to geophysical and historical factors.

The soil in this region is of two types. Land used for paddy lies about 70 feet above mean sea-level; the soil is claylike and the land is subject to frequent flooding, especially during heavy rains. In contrast, highland ranges from 100 to 377 feet above mean sea-level; the soil is composed largely of laterite (locally called *kabuk*), which is porous rock unsuitable for paddy. One may assume that the land area presently in use for paddy is all that is available within the village.

Historical factors have contributed to this pattern of land use. During the first half of this century large, commercially-run, plantations in coconut, rubber and tea were opened in this area; this affected the villagers in two ways. The poorer villagers were able to find employment in these plantations, and some continue to do so even today. Mulgama villagers with some capital also

2 The boundaries are specified in the Register of Lands and the Register of Deeds kept in the District Land Office situated in the town of Panadura (about 18 miles off Mulgama). A complexity of historical factors, some being partition and the sale of village land by the British colonial government, have led to the creation of many different plots. (See also Selvadurai 1973 for a complete account.)

opened up small-scale coconut and rubber plantations in the village itself. With the increased demand for natural rubber during the second World War, these individuals were able to increase their wealth and social status substantially, a process that contributed in part to the emergence of a village elite. About 300 acres in highland are now under rubber, and about 100 are under coconut. The plantations vary in size for operational purposes, ranging from only three or four coconut or rubber trees in home gardens to holdings of about 10 acres.

Since Mulgama's inception its acreage under paddy had gradually expanded. According to the earliest written record, a Dutch land register (*thombu*) compiled for Mulgama in the 1770s, the paddy acreage at that time was about 65.3 acres, which has increased to today's 163 acres in the course of about two centuries.

However, the growth of the village population within the same period has outstripped the increase in natural resources. The population register accompanying the land register of 1770 gave the total village population at that time as 23 persons, a number which has increased to 1963 persons today (see Table I). In 1770, there were about 2.7 acres per person while today there are only a mere 0.3 acres per person.

Table I
Population of Mulgama from 1911 to 1971

Year	*No. of Persons*
1911*	714
1921*	879
1931*	965
1941+	1205
1951+	1450
1971+	1963

* Source: Town and Village Statistics, Census Publications of Ceylon.
\+ Source: Householders' Lists, Divisional Revenue Office, Horana, Sri Lanka.

It is likely that similar demographic pressures prevail in other villages in the area. Although data on population for individual villages are unavailable, Table II indicates how population has rapidly increased in the Kalutara District as a whole.

Table II
*Population of Kalutara District from 1871 to 1946**

Year	*Population*	*Numerical Increase*	*% Increase*
1871	145,741	—	—
1881	165,021	19,280	13.2
1891	183,632	18,611	11.3
1901	229,857	46,225	25.2
1911	279,493	49,636	21.6
1921	323,704	44,211	15.8
1931	362,785	40,081	12.4
1946	456,572	92,787	25.5

* Source: Ceylon Census and Statistics, Department of Census, Ceylon, 1946.

Land and Kinship Structure

I will delineate the salient features of Mulgama's kinship structure and indicate the role of in-migration in contributing to a reduced land-and-person ratio. Attention will also be drawn to the "traditional" methods by which certain male in-migrants integrated with the host community.

With the exception of four individuals, who work and live in Mulgama as household servants (2 persons), and who are employed in non-agricultural pursuits (1 is a rubber-tapper, and another has opened up a laundry), all Mulgama people are kinsmen through ties of affinity or consanguinity. Yet, all are not equals in kinship status; there are important conceptual boundaries which demarcate different degrees of closeness and distance between relatives, and which give a certain heterogeneity to the village.

First, there is a category of persons, about 927 in number, who call themselves Mulgama's "citizens" (*game minissu*), and who trace descent from a man named Loku Appu, whom they regard as the founder[3] of the village. All citizens claim patrilineal descent from Loku Appu, and they signify this by using his surname (*samarakoon achige*), which is prefixed to an individual's personal name. Villagers attach a patrilineal value to a surname, which a child inherits from his or her father, and not from the mother; a woman retains her surname even after marriage, and her children inherit the surname of her husband. Informants also said that "all people who have the same surname have the same blood".

About three-fourth of the entire body of citizens (or 711 out of 927) are "naturalized citizens" (my term); the latter have descended from early, male in-migrants, who, after marrying Mulgama women, adopted Loku Appu's surname. Although naturalized citizens are affines of the agnatic core, they became fictive blood relatives by resorting to the strategy of surname-adoption. The practice is not unique to Mulgama; Sir Henry Maine, in his studies of the Indian village communities, noted its widespread character, and Obeyesekere (1967) has observed its prevalence in a Sinhalese village in Sri Lanka's Southern Province.

Thirdly, there is a category of persons (about 940 in number), whom Mulgama's citizens refer to as "outsiders" (*pita minissu*); the term refers to all in-migrants who did not adopt the surname, to their descendants, and to females who have resided virilocally.

---

3 The recognition of Loku Appu as the founder may be premised on the fact that he held certain lands in Mulgama from the Dutch colonial government (1656–1796) as payment for his services as the regional village headman *(arachi)*. Though others in the village held land on rent, the former was the only holder of heritable land: "Under the Dutch the service holdings are either accommodesan, that is land given to an individual while holding office and as remuneration therefore, or paraveni divel. The first was held by chiefs, the second by smaller headmen and others... Accommodesan was neither heritable nor alienable. Paraveni divel, as the name denotes, were heritable subject to due performance of the service for which they are held" (Codrington 1938:19).

Thus, Mulgama's kinship structure consists of an inner circle or core of real citizens, who are patrilineally descended from the founder, and who are defined by two categories of affinal kin: a covert category of naturalized citizens, and the outsiders whose affinal ties with the agnatic core are explicit.

This (emic) set of kin categories has a direct bearing on certain modes of resource allocation prevalent in Mulgama, today. It increases in significance when one considers the numbers of persons who come under each category. It will be noted that the agnatic core, excluding (196) females who reside virilocally, forms only about 10% of the total population. It is not surprising that a great many individuals in the remaining 90%, who had previously enjoyed citizenship and land rights in an equal manner with agnates, have become the object of discriminatory actions of certain Mulgama people. As long as there was enough land the agnates could host in-migrants and make room for their progeny, and it is likely that the conceptually demarcated distances among kin also stand for methods and strategies of disenfranchising certain kin so that others may gain.

Ownership Holdings in Paddy

A scrutiny of paddyland holdings of individual households offers further perspectives on the effects of population increase. As indicated in Table III, a significant number (44% or 11 out of 25 households) own less than half an acre per household, this being largely an outcome of subdivision of shares at each new generation and among increasing numbers of co-parceners. Since, by village standards, only about half an acre of paddy can provide the subsistence requirements of the average village household, these families cannot and do not depend solely on their shares in paddy to make a living: out of the 11 household heads, 2 are agricultural laborers, 4 are sharecroppers, 1 is a mason, another is a carpenter, 2 are road-construction laborers, and the last is a rubber-tapper. They also depend heavily on government-subsidized rice, although every family in the village also buys it at the Mulgama co-operative store.

Table III
*Ownership Holdings in Paddy for 25 Mulgama Households**

Size (in acres)	*No. of Households Owning*	%
.25– .5	11	44
.6 –1.0	6	24
1.1 –1.5	5	20
2.0 –8.25	3	12
Total	25	100

* I selected 25 households to represent the different socioeconomic levels in the village, being guided in their selection by the willingness of household heads to give me accurate information on a 'sensitive' subject, namely, land. I also tried to include representatives of the various kin categories in the sample.

Other Mulgama families enjoy a less precarious position. Although, at the present time, there are 24% of them owning between half and one acre per household, it is likely that, with the subdivision of the parental holdings in subsequent generations, many more will join the ranks of persons for whom paddyland has ceased to provide the minimum subsistence requirements.

More disconcerting, perhaps, is the concentration of a relatively large acreage in the hands of a few Mulgama people. Note that, in Table III, there are 3 individuals owning between 2 and 8.25 acres per household; one (a schoolteacher) owns 2 acres, another (also a schoolteacher) owns 4.25 acres, and the third (a son of the former village headman) owns 8.25 acres. Altogether they own a paddyland area (14.5 acres) which roughly equals that owned by the other households in the sample (i.e., 15.9 acres).

It would be incorrect to infer from the pattern of ownership holdings that the land-owning elites are an oppressive class as such. Certainly, as I will indicate in the next part of the paper, some of them gained land at the expense of the non-elites. Also, on account of their present wealth and influence the elites have a superior position in village affairs, and have ready access to such institutions outside the village as the local police and government offices located in the neighboring towns. At the same time, the non-elites need them to obtain loans, to secure jobs for their children, and to get advice on such matters as ligitation in which many villagers engage. The elites are also technological innovators, particularly with respect to paddy cultivation, and villagers benefit from their innovations. For example, the two threshing machines, each owned by the two schoolteachers, are widely used in the village, and have filled the technological gap left by a declining cattle population. The elites have also played the role of ideological change agents by emphasizing hierarchic values in their strategies to gain more land, and have provided the non-elites with models for emulation.

PART II. STRUCTURAL CHANGE

The rapid increase of the village population forms the backdrop for the change in kinship and property rights from egalitarian to hierarchic values. This increase led certain individuals to adopt new strategies in the struggle over diminished resources. The focus, here, is on one such strategy, namely, that of allocating land on a differential basis among different kinds of kinsmen. Other strategies have traditionally been available to villagers who had to cope with similar problems in the past, although population pressures were not as acute then as they are today. One way of presenting the ideological change is by contrasting the traditional methods with the new, and, second, by examining the role of individuals who brought the change about.

Traditional Strategies

Polyandry, cross-cousin marriage, inter-village migration, and certain land tenure arrangements for dividing the land and/or its yield among co-parceners, are some of the traditional methods of alleviating population pressures. Although many of these practices have declined in importance today, villagers, particularly the non-elites, had recourse to them in the past; the elites seem to have preferred other avenues even then.

Polyandry

A polyandrous union, a traditional form of family planning, involves two male siblings who share one spouse, and who live with her children under one roof. The men pool their shares in land and cultivate them as an undivided unit. In this way a sibling group manages to prevent subdivision of the parental property, and to produce fewer heirs to their own estate.

The data indicate that polyandry was not widely practised among villagers, for, in the course of Mulgama's two-hundred year history, there have been only 9 such unions, a large number (7) of them taking place during the second half of the nineteenth century. Several factors may account for the decline of this practice in recent times. Villagers consider it shameful, and even the two men, who maintain this arrangement today, were reluctant to divulge it. These attitudes among non-elites may also reflect their attempts to emulate the village elites, none of whom have maintained a polyandrous arrangement.

Cross-cousin Marriage

Since people inherit land along direct, biological lines (from parents to their biological offspring), cross-cousin marriage, as an individual strategy for obtaining more land, involves a man's marriage to his *real* father's sister's or his mother's brother's daughter. By marrying his patrilateral cross-cousin a man adds his father's sister's share to his own in the parental property; he makes a similar gain by marrying his matrilateral cross-cousin: he adds his wife's share in her parental property to that of his mother.

In spite of its strategic advantages cross-cousin marriage has not been a widespread practice, since only eight of these marriages have taken place in Mulgama's entire history. In a majority of instances (5), the marrying partners were Mulgama residents prior to marriage, and it is clear that these marriages were attempts to consolidate existing affinal relationships between Mulgama families.

Why have so few cross-cousin marriages taken place despite the fact that, according to informants, the cross-cousin (*avassa nena*, the term for the female relative) is the most suitable marriage partner (the term *avassa*, a derivative of *avasya* meaning "necessary" may also denote the importance of marrying a cross-cousin in the context of a land shortage)? The answer partly lies in the fact that, although the marriage is a solution to the problems of one man, it creates problems for others, namely, other co-parceners who may be enjoying the shares which the former now claims on the basis of marriage. These conflicts of interest may also account for continuing animosities among brothers and between men and their affinal kin once a cross-cousin marriage has taken place. For example, there is an on-going dispute between two brothers over the share claimed by one through his marriage to the father's sister's daughter. Another dispute has arisen between a man and his mother's brother's sons over the former's claims to the shares of his wife and his mother. In other instances, even though animosities are not overtly displayed, the men claiming these shares are distrustful and suspicious of their affinal kin; and many convey these feelings in terms of the idiom of supernaturally-caused illness (*dos*), which, in their perception, accounts for most misfortunes befalling them or others in their households.

Migration

Inter-village migration as a method of distributing excess population in areas with excess land has been a recurrent feature of Sinhalese peasant society. Although in certain regions, particularly in Sri Lanka's Dry Zone, entire villages or groups of kin may migrate in search of cultivable and unoccupied land, in the Wet-Zone with its settled villages relying on a regular water supply and supporting a dense population, the unit of migration is usually the individual. Since a person almost always migrates from his natal village on marriage, he requires the support of his close kin, who may be assisted at times by professional marriage brokers (found in every village), to arrange a marriage. When a person leaves his kin on marriage he or she goes with the objective of residing permanently in another village; separations are uncommon (4 out of a total of 432 marriages taking place within the past three generations).

A group of siblings stands to gain by sending out some of its members, especially when there is not enough land for all; the sum of these out-migrations also means more land per person for the village as a whole. A scrutiny of marriages, which have taken place within the past three Mulgama generations brings out more clearly the nature of these individual strategies considered *ensemblé*, and their relation to population pressures on land.

Table IV
Numbers of Persons Out-Migrating and Remaining Resident Within the Past Three Mulgama Generations

	Males	*Females*	*Total*	%
No. Out-Migrating	48	109	157	36
No. Remaining Resident	186	89	275	64
Total	234	198	432	100

As indicated in Table IV, about a third (36%) of all persons born during the period considered have left Mulgama on marriage, with 6 persons settling down in a village expansion scheme adjacent to Mulgama, 22 persons leaving for Colombo and its suburbs, and the remaining 129 persons going out to live in other villages. Although a large majority of out-migrants (109 out of a total of 157) were women, a fair number (89 women) also remained resident; this includes 11 unmarried girls (all between the ages of 15 and 20); 5 confirmed spinsters; 46 who married men of Mulgama; and the remaining 27 who married in-migrating men. Apparently some Mulgama families had enough land to share with fellow villagers and with migrants from other villages. It is possible that some have remained unmarried and others have postponed early marriage due to population pressures in other villages.

A sibling group or a village does not solve all its problems by simply sending out its excess membership, since spouses have to be brought in for those remaining resident. Thus, a fundamental aspect of marriage strategies is to ensure that the incoming number is less than, and certainly not greater than, that of outgoing persons. And, if the incoming and outgoing numbers are the same, the person/land ration would remain unaltered. How does each sibling group and in turn each village strive to offset the balance in its favor?

Mulgama villagers maintain to a degree a system of unbalanced reciprocity through an extensive rather than an intensive marriage network. A total of 177 marriages have taken place between Mulgama families and those in 82 other villages, all situated within a radius of 21 miles from Mulgama (not including a number of other marriages for which data are not available). This means that on the average roughly 2 marriages have taken place between Mulgama and another village, with the excess population of each being thinly spread over several others. The low density of the marriage network may also indicate land shortages in most villages in the area.

Another important feature is the non-reciprocal character of the marriage relationship between Mulgama and certain villages. For example, although 46 persons have out-migrated from the former into 36 other villages, none have emigrated from the latter into Mulgama, and another 24 persons have in-migrated from villages that have not taken in any from Mulgama. We may assume that in these instances marriages have taken place between non-kin, this being a strategy radically different from that underlying cross-cousin marriage.

The physical distance between villages also plays an important role in maintaining a system of unbalanced reciprocity. It will be noted in Table V that Mulgama families have sent out more members and received a fewer number from the more distant villages, the network being more extensive and probably more advantageous to the former with villages neither too close nor too far (i.e. beyond two and less than seven miles from Mulgama). On the other hand, there is a tendency for a greater number of persons to in-migrate from villages in the immediate vicinity of Mulgama.

How does a villager gain by sending out certain siblings to more distant

Table V
Villages Forming a Marriage Network with Mulgama

Distance (in miles) from Mulgama	*No. of Villages**	*No. of Persons Migrating out of Mulgama*	*No. of Persons Migrating into Mulgama*
0–2	12	28	40
2–7	39	50	27
7–20	21	22	10
Total	72	100	77

* There were 10 villages whose location or distance from Mulgama could not be established.

villages or by marrying them out to non-kin? Though many of these villages are not inaccessible, the farthest being reachable during a day's journey by bus, the separation from close kin and the prospect of living with strangers involves considerable emotional hardship, at least initially, for the out-going persons. Deprived of direct kin support from their natal village they may come to depend more and more on their affines. Kinship ties are not lost in this way, but, as one informant stated, "they fall by the wayside", and the estrangement is in a sense advantageous to the siblings remaining resident. The latter will divide among themselves and use the out-migrants' shares; they hope that eventually the latter would lose interest in the parental property, and after three or four generations the descendants of those remaining resident will even ignore these claims. By the same token, relationships with siblings who have gone to reside in villages closest to Mulgama tend to continue through time; gifts are exchanged and resident siblings will share the harvest with out-migrants.

Land Tenure

It is not possible to give an exhaustive discussion of land tenure arrangements within the limits of this paper. Some practices which Mulgama people have developed to cope with the needs of an increasing population may be briefly indicated as follows:

1. Rotation of shareholders (known as *katti maru*, from *katti* to mean a group of persons or a section of land, and *maru* to mean change or taking turns). Under this form of tenure many co-parceners have divided themselves into groups, and the paddyland into a fixed number of sections. Since the number of groups is always greater than the number of sections, a certain number wait (for one season, and sometimes a longer period) for their turn to cultivate a section. With each descending generation the groups subdivide, thereby making the waiting period longer; for example, one Mulgama man has to wait 17 years to cultivate his share, this being the longest wait to my knowledge. Thus,

certain people do not cultivate during certain seasons or years, although in theory they "own" paddyland.

2. Division of the yield. A group of siblings owning very little land (usually less than half an acre) divide the yield instead of cultivating this area separately. The person who does the cultivating (a sibling or, sometimes, a sharecropper) gets half the harvest plus his share of the remaining half if he is a co-parcener, and the others get their share of the remaining portion.

3. Division of proceeds from leasing shares. Instead of cultivating or renting out their shares a group of siblings will lease out their shares and divide the cash among themselves. This practice is rare (only one instance being observed) since it means that the former will have to depend on other means of gaining a subsistence.

The increasing subdivision of land is to some extent an inevitable outcome of the natural increase of the village population. Each new generation faces the prospect of reduced shares, longer waiting periods to cultivate them, and lowered yields. Certain individuals attempt to counter this process by purchasing the shares of non-resident co-parceners in order to enlarge their own shares. However, other resident co-parceners do not welcome such actions since it means a reduction of the area they cultivate. This may account for the relatively few number of people who have purchased shares (4 in a sample of 25 household heads); all have faced combined opposition from other co-parceners in attempting to claim these shares.

New Strategies and Village Elites

The avenues which the non-elites traditionally use are premised on the value of sharing, even if this means sharing less and less of the same good. Within this ideological framework various strategies and counter-strategies take place with each individual, each sibling group and each village turning the balance, temporarily at least, in its favor. No one strategy is a perfect solution; each is countered and qualified by other strategies. I will now discuss a new strategy which the non-elites have incorporated into their repertoire of traditional ones. The hierarchic values underlying the former are largely an outcome of the rise of local elites and their role in the arena of village politics. Mulgama elites are not an exception; Obeyesekere (1967) has noted the evolution of land-owning status groups called *pelantiyas* operating on an inter-village basis, and Sarkar and Thambiah (1957) have assessed their impact on landlessness. I regard these elites as agents of structural change within the village community itself since non-elites have adopted and emulated their strategies.

A constellation of economic and political factors contributed to the emergence of a landed elite in Mulgama. They were a genealogical segment of the citizenry, and their rise to prominence began with one genealogical ancestor who held three village-level offices under the British colonial government, namely, Police Sergeant from 1890 to 1900, Village Headman from 1895 to 1920 and Divisional Village Headman (*arachi*) from 1920 to 1930. One of his sons succeeded to the village headmanship in 1920; another son, whom I will call Baronchi, took over the position in 1935, and rose rapidly to a position of political and economic dominance in village affairs; I will be concerned primarily with his career.

The village headmanship gave its holder the power to settle disputes, to collect taxes on behalf of the colonial government and to maintain peace and order; the headman also had a large say in all administrative dealings which the central government had with the village. In addition to these privileges, Baronchi had accumulated a relatively large amount of cash by operating a graphite mine in the vicinity of Mulgama, and, by the late forties, he had enough capital to open up a small rubber plantation on land he inherited from his father.

One of the more prominent aspects of Baronchi's career was "landgrabbing", which was not a simple case of brigandage since all informants stated that he only grabbed land in dispute. Although as a part of his duties as headman he could intervene in disputes, he went further by persuading or forcing the disputants to sell him the land. Having the cash to do so he had amassed about 25 acres in highland and 15 acres in paddy.

It is important to note that Baronchi exploited a situation which was open to differences in interpretation, the disputes being a reflection of a normative inconsistency. Informants stated that, in a large majority of instances, the disputes had been between naturalized and real citizens over the claims of the former to the full shares of local women whom their ancestors had married long ago. It is probable that more and more citizens were unwilling to recognize these claims as population pressure on land increased. The normative inconsistency lay in the fact that naturalization was no longer a guarantee of equal rights in Mulgama land, although certain in-migrants had made a commitment to the village community by adopting citizenship. In the course of fieldwork I came across four household heads who had lost a large part of their shares in paddyland to Baronchi: 3 of them were naturalized citizens and the fourth was an outsider.

Since Baronchi's demise other elites appearing on the Mulgama scene have adopted a similar strategy. Among them are two schoolteachers who have used their influence to hold a crucial position with respect to paddyland, namely, that of "chief arbitrator" (*nayaka*) who meets with co-parceners once a year to decide on the portion of paddyland each would cultivate, and to settle disputes among them. Each schoolteacher acts as chief arbitrator for two groups of co-parceners: the patrilineal descendants of the founding ancestor, and the descendants of two naturalized citizens (brothers) who had come to Mulgama subsequently. Since these two groups have rights to about 80% of the total acreage, their decisions affect nearly everyone else although the latter have no say in the deliberations. Informants stated that the two schoolteachers have used their office to jeopardize the claims of certain naturalized citizens and outsiders by giving them less than the shares they claimed, and, by so doing, have increased their own shares.

Changes in Kinship Structure

Whilst strengthening their own wealth and status the elites have provided non-elites with models for re-structuring traditional kinship and property concepts.

One change in kinship ideology relates to surname-adoption, the customary method of incorporating in-migrant males, and of maintaining the fiction that the village is a group of consanguineally related kin. The data indicate that this practice has declined in the course of the past 5 generations; although there were 57 in-migrant males during this period only one adopted the surname, he too being a resident of another village close by (three of his sons came later to live in Mulgama). Prior to that all in-migrating males except one had followed the practice. Today, villagers consider surname-adoption demeaning; they regard the descendants of the one recent adoptor as "inferior" (*adui*, a term also used to refer to a person lower in caste) and exclude them from marriage ceremonies which involve commensality.

The emphasis on hierarchy as an ideological strategy for displacing certain persons from the moral community has also permeated the body of the citizens, as well. This has taken the form of a rank order of statuses and rights depending on their relationship to the founding ancestor. According to one informant there are five sub-categories of citizens: the "first" (*palaveni*, also meaning the first to arrive) are the ancestor's patrilineal descendants, who also use his full surname (*egodahage nohoth samarakoon achige*); the "second" (*deveni*), "third" (*tunveni*) and "fourth" (*hataraveni*) classes consist of the naturalized citizens who use only a portion of the surname (the last two words), and whose rank order is determined by length of residence, those residing for a longer period being superior to later arrivals. According to him the lowest in the rank order are some virilocally resident women, who have defied the traditional patrilineal ideology by adopting the surname of their husbands.

Changes in Land Rights

Hierarchic concepts are also used in Mulgama to re-define the land rights of naturalized citizens and outsiders, whose claims (through affinal ties) are referred to as *adui*, that is, unequal to or less than those of agnates. This refers to a system of differential allocation of rights on the basis of sex prevalent in Mulgama today; under this system the villagers consider the use rights (*buktiya*) of females as being subordinate to those of their agnates, and, in fact, the term "female inheritance" (*ganu urume*) is used to denote the bundle of rights in land to which a woman, and, by extension, all affines are entitled.

The system of differential allocation of use rights takes several different forms today. For example, in a sample of 9 household heads who were outsiders, only two had access to the full shares of females; one had benefited from partition proceedings which a co-parcener, a citizen, had effected through the local courts of law, and the other had used his influence as the secretary of the Mulgama Cultivation Committee. Three other household heads had full shares in paddy, though not in highland, and the remaining four had full shares in highland and not in paddy.

The disenfranchising process has also led to disputes over land; a large majority of on-going disputes (6 out of 9)[4] are between citizens and others (i.e., outsiders and naturalized citizens), who have challenged the system of unequal allocation[5]. These disputes are long-drawn out, and some outsiders have resorted to litigation as a way out of the problem.

Conclusion

My objectives were to describe and analyze changes in kinship and land rights in a Sinhalese village community, and to view these changes in the light of demographic trends. I have presented the former in terms of a shift from egalitarian to hierarchic values, which have led to an ideological re-structuring. The symptoms of this change process have been the intensification of (conceptual) boundaries between agnates and non-agnates, between citizens and non-citizens, and the emergence of a system of differential allocation of land rights emphasizing the claims of men over women, and agnatic over affinal ties. Another result is the displacement of some, if not all, naturalized citizens, who, in the past, had enjoyed the rights and privileges of citizens. These changes have also led to some degrees of dissension and conflict as evidenced by land disputes, litigation and a general feeling of disenchantment within the ranks of the disenfranchised.

Though the shift in values is not sharp and irrevocable for some villagers, who continue to use traditional along with the new strategies premised on hierarchy, it is likely that the former are on the decline. Such traditional strategies as cross-cousin marriage and surname-adoption were meaningful within a certain land-person ratio which does not obtain today as a result of rapid population increase in almost every village in the locality. A traditional strategy having limited importance now is inter-village migration; the current trend in out-migration is to the cities and towns. Many Mulgama migrants to Colombo have gone to live with kinsmen who had established themselves in the

4 The greater number of disputes over highland may be related to the fact that the use of highland for residence and for the cultivation of garden trees such as jak, rubber and coconut, which are known as 'permanent property' *(isthira depala)*, symbolize permanent occupation of the land, and thereby citizenship in a certain sense. In contrast, paddyland is used to cultivate a 'temporary' crop lasting 3 or 4 months.

5 The claim for equality of shares of both males and females is based on the law upheld by the courts, namely, Roman-Dutch law. In a separate paper (Selvadurai n.d.), I have examined the impact of the law of courts on these issues over land rights in Mulgama. In a sense, the law and the courts counteract the trend towards hierarchy by upholding a principle of equality before the law.

urban setting, and it is likely that villagers are giving a greater importance to kinship ties with urban dwellers rather than with their rural counterparts.

The decline of traditional values does not signify a state of normlessness within the village itself, although some individuals by emulating the local elites have gained at the expense of other villagers. Rather, we find a re-integration of the moral community in terms of an "old" principle. As Obeyesekere (1966, 1967) has shown, the "new" hierarchic principle, which has re-defined kinship and land rights, is not entirely alien to the Sinhalese villager, his caste ideology, his historical experience and his very belief structure being modeled on hierarchic values.

In spite of its congruity with the larger conceptual scheme of traditional Sri Lanka, the emergence of hierarchy in the realm of kinship and land rights remains, like all other solutions, an imperfect one; although it solved the problems of some villagers, it created problems for others. Two young persons from Mulgama (a male and a female) participated actively in the April, 1971 insurgency. Both came from families with very little land and with poor incomes; these families had also been disenfranchised through the new emphasis on hierarchy.

REFERENCES

Arulpragasam, L. C.
1961 "A consideration of the problems arising from the size and sub-division of paddy holdings and the principles and provisions of the Paddy Lands Act to them". *Ceylon Journal of Historical and Social Studies,* 4:59–70.

Codrington, H. W.
1938 *Ancient Land Tenure and Revenue in Ceylon.* Colombo: Ceylon Government Press.

Leach, E. R.
1967 *Political Systems of Highland Burma.* Boston: Beacon Press.

Obeyesekere, G.
1966 "The Buddhist pantheon in Ceylon and its extensions", *Anthropological Studies in Theravada Buddhism,* Yale Southeast Asia Studies, Cultural Report Series No. 13.
1967 *Land Tenure in Village Ceylon: A Sociological and Historical Study.* Cambridge: Cambridge University Press.

Perera, P. D. A.
1968 "Some Agricultural Policies and their Effects on Peasant Agriculture of the Wet Zone of Ceylon during the Nineteenth Century", *Vidyodhava Journal of Arts, Science and Letters,* 1:37–52.

Sarkar, N. K.
1957 *The Demography of Ceylon.* Colombo: Government Press.

Sarkar, N. K. and Tambiah, S. J.
1957 *The Disintegrating Village.* Colombo: University of Ceylon Press Board.

Selvadurai, A. J.
1973 *Culture and Continuity: A Study of Kinship and Land Tenure in a Sinhalese Village.* University of Washington: Unpublished Ph.D. dissertation.
n.d. "Land, personhood and sorcery in a Sinhalese village", in B. Smith (ed.), *Journal of Asian and African Studies* (special issue). Forthcoming.

CONTRIBUTORS

Dr. Lucien M HANKS was a senior Research Associate at Cornell University, and a member of the faculty of Social Sciences. He is now retired and lives at North Bennington, Vermont.

Steven and Eleanor PIKER have conducted research in Thailand during 1962–63 and 1967–68, both times with the support of NIMH. The research focused on social organization, child rearing, and Thai religion. He has authored a number of publications dealing with these topics, beginning in 1966. Steven Piker's Ph.D. in Anthropology is from the University of Washington (Seattle).

David JOHNSTON is an instructor in history at the Phillips Exeter Academy in New Hampshire. He has done graduate work at the University of Washington and at Yale, where he is currently finishing his dissertation for a Ph.D. in history. His dissertation, "Rural Thai Society and the Development of the Rice Economy, 1880–1930," is based on research conducted in Thailand during 1972 and 1973.

Charles F. KEYES carried out anthropological research in Thailand in 1962–1964, 1967–1968, and 1972–1974 and served in 1972–1974 as Fulbright Exchange Lecturer and Advisor at Chiang Mai University. His publications include: *Isan: Regionalism in Northeastern Thailand;* "Buddhism and National Integration in Thailand;" "The Northeastern Thai Village: Stable Order and Changing World;" "Kin Groups in a Thai-Lao Village;" and "Local Leadership in Rural Thailand."

Patrick PEEBLES is an assistant professor of history at the University of Missouri-Kansas City, giving courses on Asian history and quantitative methods for historians. Before joining UMKC he was visiting assistant professor of history at the University of British Columbia and at the University of Michigan. He is working on a monograph examining the social origins of the modern political elite of Sri Lanka in the period from 1810 to 1930. He received his Ph.D. from the University of Chicago in 1973.

James BROW has been assistant professor of Anthropology at Swarthmore College since 1971. He obtained his Ph.D. from the University of Washington in 1974. His dissertation on the history and social structure of Veddah villages in Anuradhapura District, Sri Lanka, was based on two years field research in the island between 1968 and 1970.

Anthony SELVADURAI was born December 24th, 1940 in Sri Lanka and received his graduate training in anthropology and his Ph.D. (1973) at the University of Washington, Seattle. His dissertation topic was land tenure, land disputes and kinship structure in a Sinhalese village in Sri Lanka's Western Province (1971–72) and his interests remain in this field. His further interests are symbolic anthropology, social change through Western law in South Asia, and economic development.